FUNDAMENTALS OF NUCLEAR SCIENCE

FUNDAMENTALS OF NUCLEAR SCIENCE

With Applications in Agriculture and Biology

P. N. TIWARI
Indian Agricultural Research Institute
New Delhi

A HALSTED PRESS BOOK

JOHN WILEY & SONS
New York—Toronto

*Published in the U. S. A., Canada
and Latin America by Halsted Press,
a Division of John Wiley & Sons, Inc.,
New York*

Library of Congress Cataloging in Publication Data

*Tiwari, P. N.
Fundamentals of nuclear science, with applications in
agriculture and biology*

*"A Halsted Press book."
1. Nuclear physics. 2. Atomic energy. 3. Atomic
energy in agriculture. I. Title.
QC776.T58 1974 539.7 74-1290
ISBN 0-470-87522-4*

Preface

In the short span since the dawn of the nuclear age, it has become clear that nuclear science offers unique, and often indispensable opportunities for investigating the living world. Several important discoveries in biology and agriculture during the last two decades have been made possible by the application of nuclear science. The first few years of the applications of isotopes to the investigation of photosynthesis have brought about more successes than the preceding 180 year period. The use of isotopically labelled fertiliser is the only known method to measure directly the uptake from the fertiliser. Rapid and nondestructive evaluation of oil in oilcrops cannot be accomplished except through nuclear magnetic resonance technique.

At present isotopes and radiations are being used in almost every branch of agriculture and biology. Most problems associated with the uses, however, are concerned with the technique of safe handling of radioactive materials, performing proper measurements involving radioactive and stable isotopes and interpreting data obtained from the measurements. Much of the mystique associated with them can be overcome by an understanding of the fundamentals of nuclear science.

Part 1 therefore, emphasises the basic concepts of nuclear science with a minimum of mathematics. In this part the atom, nucleus, radioactivity, interaction of radiation with matter, radiation detection and measurement, and radiation protection have been discussed in a manner that they can be easily understood by biologists and agriculturists who do not have advanced training in mathematics and quantum mechanics. In fact, most of the material covered in this part was delivered as series of lectures to the agriculturists and biologists during a summer course and a three months training course in 1972. The response from the participants was encouraging.

Part 2 describes applications to agriculture and biology. These have been discussed under three categories, that is, isotopic tracer, nuclear irradiation and nuclear analytical techniques. The main effort has been devoted to explaining the underlying basic principles of application. Applications of isotopes in soil-plant nutrition, plant physiology and biochemistry and hydrology are given. Each category of application has been illustrated by taking examples from different branches of agriculture and biology. Under the category nuclear analytical techniques, a detailed account of activation analysis is given. The basic principles of neutron moisture meter and the NMR technique for rapid and nondestructive determination of oil in oil seeds have also been discussed.

In writing this book I have consulted most of the books and research articles listed in bibliography. I hope the bibliography will be useful for ambitious readers. I am thankful to Dr. N. P. Dutta, the former Project Director of Nuclear Research Laboratory, for encouraging me to write this book. I am also thankful to my wife Nirajana for her patience and cooperation while the book was in preparation.

P. N. Tiwari

New Delhi
August 1973

Contents

PART I

Chapter One

THE ATOM

Chapter Two

THE NUCLEUS

Chapter Four

INTERACTION OF NUCLEAR RADIATION WITH MATTER

Chapter Five

NUCLEAR RADIATION DETECTION AND MEASUREMENT

CHAPTER ONE

The Atom

1.1 STRUCTURE OF THE ATOM

1.1.1 J. J. Thomson's Model of the Atom

The Atom is the smallest possible particle of an element that can take part in chemical combination. Normally it contains equal amount of positive and negative charges. Therefore, the normal atom is electrically neutral. The carrier of positive charge is called proton and that of negative charge is called electron. The proton is 1836 times heavier than the electron, their charges being equal in magnitude and opposite in signs. The atom is nearly spherical (radius $\sim 10^{-8}$ cm). In 1910, J. J. Thomson proposed that the atom is made up of a number of electrons moving in a sphere of uniform positive electrification as shown in Fig. 1.1. This means that the positive charge of the atom is spread over its entire volume.

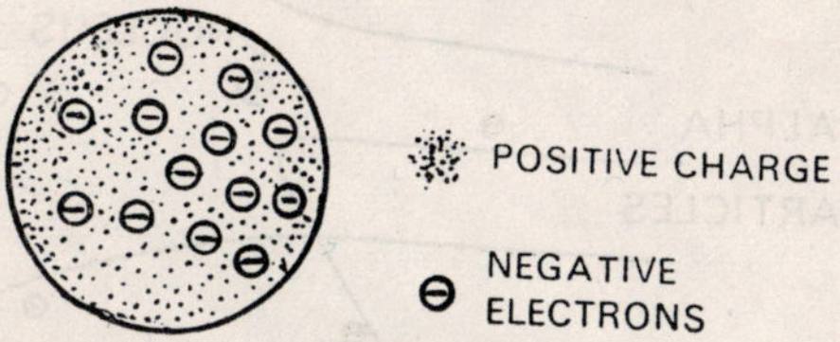

Fig 1.1 J. J. Thomson's Model of the Atom.

1.1.2 Rutherford's Model of the Atom

The modern idea about the structure of the atom arose in 1911 from the experiment of Ernest Rutherford on the scattering of alpha particle (helium nucleus) from thin sheet of metal. In this experiment a sample of polonium, a radioactive alpha-emitter is placed in a box, so that only those alpha particles which hit a small hole in the side of the box can escape. The narrow beam of escaping alpha particles is directed to a very thin foil (10^{-5} cm) as shown in Fig. 1.2. A vast majority of the particles went straight through the foil, as if it was empty, but a few of them deflected at various angles, some nearly 180°. This is contrary to the predictions of J. J. Thomson's model of the atom. In order to explain the results of his experiments Rutherford proposed that the positive charge of the atom is not uniformly spread over the whole atom, but it is concentrated in a small region called the nucleus, at the centre of the atom as shown in Fig. 1.3. According to this model positively charged alpha particles approaching the nucleus will be repelled and deflected, because similar charges repel each other.

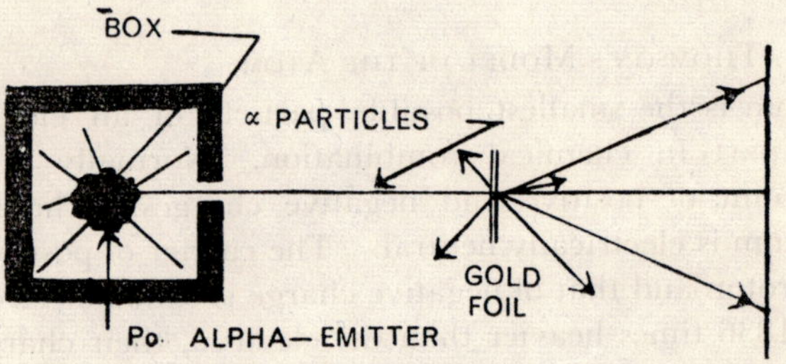

Fig. 1.2 Schematic diagram of the experiment through which the nucleus was discovered.

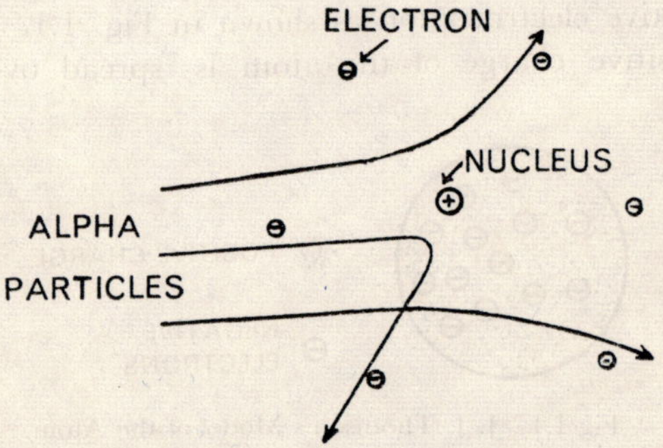

Fig. 1.3 Showing the nucleus in the centre of the atom and the scattering of alpha particles by the nucleus.

The amount of deflection would depend on how close the alpha particle came to the nucleus. This is because the electrical force of repulsion between two similar charges is proportional inversely to the square of the distance between them and directly to the product of their charges, i.e.

$$F = K \frac{Q_1 Q_2}{r^2}$$

(Coulomb's law). where F is the force between two charges Q_1 and Q_2 separated by distance r, and K is the constant of proportionality.

It is possible to get the approximate size of the nucleus by careful analysis of the number of alpha particles scattered at different angles. Rutherford measured the size of the nuclei of various elements. He found them to be only about 10^{-12} cm in radius, whereas atoms themselves are about 10^{-8} cm in radius. The volume of the atom is about 10^{12} times greater than that of the nucleus, although the nucleus accounts for practically all the weight of the atom. Thus, most of the atomic volume is relatively empty. In other words, the nucleus is enormously more dense than the atom. The density of the nucleus is about 10^{8} tons per cubic centimeter, never encountered in common experience.

1.1.3 Bohr's Model of the Atom

In Rutherford's model of the atom the protons are assumed to be concentrated in a small space, called nucleus, at the centre of the atom and electrons are outside the nucleus. To account for the fact that electrons did not fall into the positively charged nucleus due to electrical attraction between dissimilar charges, Rutherford found it necessary to propose that the electrons in the atom are not stationary but are rapidly rotating around the nucleus as the planets rotate around the sun. The inward attractive force was supposed to be balanced by the outward centrifugal force. However, the analogy between the atom and the planetary system is not correct because the particles in the atom are electrically charged. By the electromagnetic theory, the rotating electron should continuously emit energy as radiation. But if this were the case, the radius of curvature of its orbit would steadily diminish. The electron would thus follow a spiral path and eventually merge into the nucleus. Further, the atomic spectra would be continuous instead of well defined lines as observed in actual practice.

Niels Bohr calculated that the emission of radiation (electromagnetic wave) by the rotating electron would cause it to fall into the nucleus within one hundred-millionth of a second. Thus on the basis of conventional mechanics and electrodynamics, the planetary system of electrons revolving around the nucleus as suggested by Rutherford could not exist for more than an extremely short period of time. This was in direct contradiction of the fact that atoms do exist permanently and do not show any tendency to collapse. Since nature cannot be wrong, conventional mechanics must be wrong, at least, when applied to motion of electrons within an atom. Defying the well-established laws of classical mechanics and electrodynamics, Bohr in 1913 stated that in the case of the motion of electrons within the atom the following postulates are valid :

I. From all the mechanically possible circular and elliptical orbits of electrons moving around the nucleus only a few quantised orbits are permitted whose angular momentum is given by :

$$mvr = \frac{nh}{2\pi} \tag{1.1}$$

where m is the mass of the electron, v is its velocity, r is the radius of the orbit in which the electron is rotating, n is any integer and can take values $1,2,3....n$, h is called Planck's constant. That is to say that the angular momentum of the orbit is quantised and is restricted to certain values which are integral multiples, e.g. $1,2,3,...n$, of a definite quantity $h/2\pi$.

II. Circulating along these orbits around the nucleus, the electrons are prohibited from emitting any electromagnetic waves, even though conventional electrodynamics says that they should. Such orbits would be consequently stable and would represent stationary states of the atom.

III. Electrons may jump from one orbit to another, in which case the energy difference between the two states of motion is emitted in the form of single light quantum whose frequency is given by the following quantum theory equation:

$$\nu = \frac{E}{h} \tag{1.2}$$

or

$$h\nu = E \tag{1.3}$$

Where E is the energy difference between two states and ν is the frequency of emitted radiation.

If E_2 is the energy of the atom in a state of higher energy and E_1 is the value in the state of lower energy, then an electronic transition from the former to the latter state will be accompanied by the emission of energy $E_2 - E_1$. By Planck's quantum theory, the frequency of the emitted radiation would be :

$$\nu = \frac{E_2 - E_1}{h} \tag{1.4}$$

and

$$\lambda = \frac{c}{\nu} = \frac{ch}{E_2 - E_1} \tag{1.5}$$

Where λ is the wave length and c is the velocity of light. Each particular energy transition, from one state to another, should thus result in the emission of radiation with definite frequency given by equation (1.4) which results in the formation of a spectral line in the spectrum.

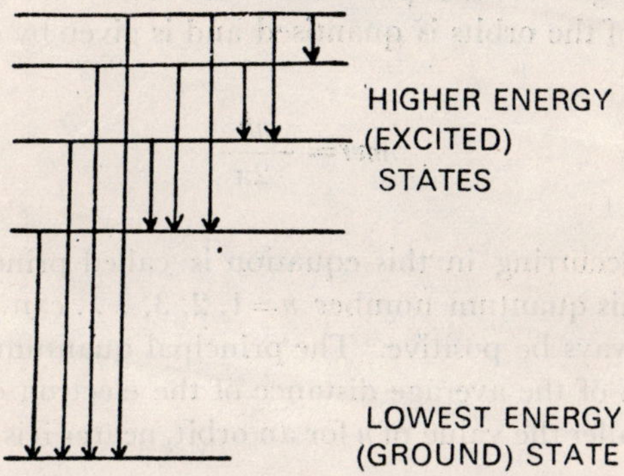

Fig. 1.4 Transitions of electrons from higher to lower energy states giving spectral lines.

The lowest energy state of the atom is called the ground state. The electrons which are normally in their lowest energy state can pass to higher energy states if energy is supplied to them by the use of high temperature or suitable electrical discharge. The higher energy states are called excited states. The electrons spontaneously return from higher to lower energy states resulting in the emission of a series of spectral lines as shown in Fig. 1.4. Since many different energy states are possible in every atom, numerous lines may be observed in the spectrum. Bohr was able to evaluate the energies for the possible stationary states of the hydrogen atom. Using equations (1.4) and (1.5) the frequencies and wave lengths of the lines in the hydrogen spectrum were determined and found to be in remarkable agreement with the experimental values.

1.2 THE QUANTUM NUMBERS

We will now discuss four quantum numbers called principal quantum number *n*, angular momentum quantum number *l*, magnetic quantum number *m*, and spin quantum number *s*. The term "quantum" specifies that these numbers must have certain discrete values and cannot assume continuous values. An acquaintance with these quantum numbers is necessary to understand the grouping of electrons into the atomic shells.

1.2.1 Principal Quantum Number *n*

According to the first postulate of the Bohr's theory the angular momentum of the orbits is quantised and is given by equation (1.1)

$$mvr = \frac{nh}{2\pi}$$

The *n* occurring in this equation is called principal quantum number. This quantum number $n=1, 2, 3, \ldots$ can never be zero and must always be positive. The principal quantum number gives a rough idea of the average distance of the electron orbits from the nucleus; smaller the value of *n* for an orbit, nearer it is to the nucleus.

1.2.2 Angular Momentum Quantum Number *l*

The angular momentum quantum number, like *n*, is related to the angular momentum of the electron due to its motion round the nucleus. For each value of the principal quantum number *n*, only certain values of the angular momentum quantum number are permitted; these vary by steps of unity from zero to $n-1$. Thus if *n* is 4, then *l* could be 0, 1, 2, 3 but nothing else.

1.2.3 Magnetic Quantum Number *m*

This quantum number was introduced to account for the behaviour of spectra in magnetic fields. It can take any integral value, including zero, ranging from $-l$ to $+l$, that is for $l=3$, $m=3$, 2, 1, 0, -1, -2, -3.

1.2.4 Spin Quantum Number *s*

This quantum number had to be postulated to account for the close grouping of two, three or more spectral lines. Every electron possesses an angular momentum due to its spinning motion. In a magnetic field the spin axis can have two directions relative to the field. This quantum number can have one of the two values : $+1/2$ or $-1/2$. The magnitude of 1/2 for both cases shows that the electrons always spin at the same rate. The plus and minus signs refer to the parallel and antiparallel direction of electron spin with respect to the applied magnetic field.

1.3 PAULI EXCLUSION PRINCIPLE AND ATOMIC SHELLS

1.3.1 Pauli Exclusion Principle

The exclusion principle, discovered by W. Pauli, states that no two electrons in an atom may have same values for all the four quantum members. In other words, no more than one electron can occupy each possible energy state.

1.3.2 Atomic Shells

Traditionally the $n=1$ electrons are said to belong to the *K* shell, $n=2$ electrons to the *L* shell, $n=3$ electrons to the *M* shell, $n=4$ electrons to the *N* shell and so on. Similarly, electrons belonging to the angular momentum quantum number $l=0, 1, 2, 3, 4, 5$ are called *s, p, d, f, g, h* electrons respectively. Pauli exclusion principle in conjunction with the rules which give the permitted values of the quantum numbers *n, l, m* and *s* can predict how many maximum electrons can be accommodated in each shell. Consider, first the shell for which the principal quantum number $n=1$, the only possible values of *n, l, m* and *s* which satisfy the properties of the quantum numbers as well as the exclusion principle are then as follows :

n	1	1
l	0	0
m	0	0
s	$+1/2$	$-1/2$

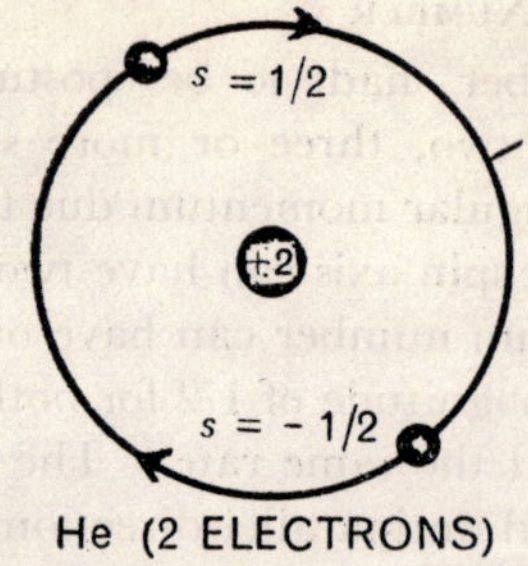

Fig. 1.5 Only two electrons can stay in *K* shell.

Consequently, there can be no more than two electrons in the *K* shell ($n=1$) as illustrated in Fig. 1.5. Since for both these electrons $n=1$ and $l=0$, they are called 1 *s* electrons. In the *L* shell with $n=2$, there are eight possibilities; thus :

n	2	2	2	2	2	2	2	2
l	0	0	1	1	1	1	1	1
m	0	0	-1	-1	0	0	1	1
s	$+1/2$	$-1/2$	$+1/2$	$-1/2$	$+1/2$	$-1/2$	$+1/2$	$-1/2$

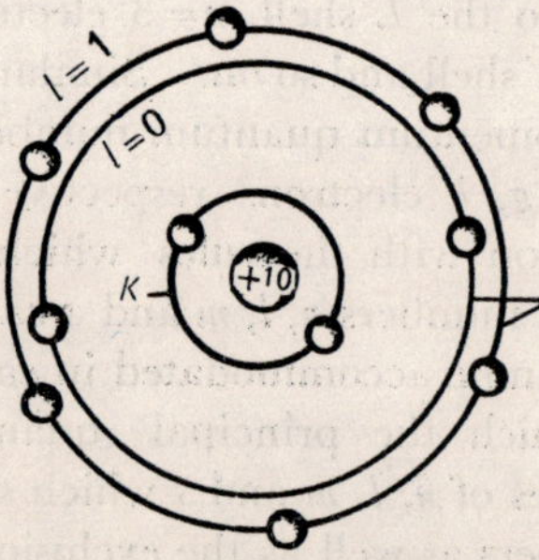

Fig. 1.6 Eight electrons can stay in L shell.

It can be noticed that in no two columns all the four quantum numbers are identical, so that there can be up to eight electrons in the *L* shell ($n=2$) as shown in Fig. 1.6. Of these eight electrons, two are 2*s* electrons with $n=2$ and $l=0$ and six are 2*p* electrons with $n=2$ and $l=1$. Similarly, it can be shown that *M* shell with $n=3$

can accommodate 18 electrons, *N* shell with $n=4$, 32 electrons and *O* shell with $n=5$, 50 electrons. It may be noticed that the maximum number of electrons in the successive principal quantum (n) levels are 2, 8, 18, 32, and 50, which may be written as 2×1^2, 2×2^2, 2×3^2, 2×4^2 and 2×5^2 respectively. From this, it can be concluded that the maximum number of electrons, corresponding to a shell with principal quantum number n is given by $2n^2$.

1.4 ELECTROMAGNETIC RADIATIONS

The combination of electrical and magnetic wave is called electromagnetic radiations. Light is the electromagnetic radiation caused by the electric and magnetic changes in the atom. White light is a mixture of all visible colours; they are essentially red, orange, yellow, green, blue and violet. Of these, red has the longest and violet the shortest wave length. The question may be asked: are there electromagnetic radiations having wave lengths both longer and shorter than visible light? The answer to this question is that such radiations are known. In the order of decreasing wave length or increasing frequency they are : power, radio-waves, microwaves, infrared, visible light, ultraviolet, X-rays and gamma rays—as illustrated in Fig. 1.7.

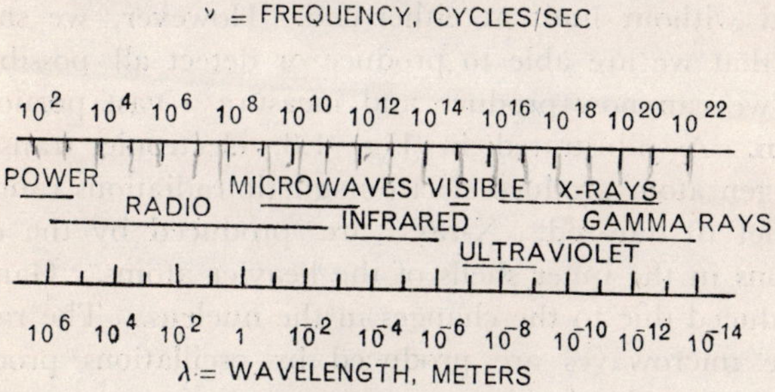

Fig. 1.7 Approximate frequency and wave length regions of various electromagnetic radiations.

These radiations are fundamentally the same; they all travel with the same speed (the speed of light) and differ only in their wave length from 10^{-14} m (10^{-12} cm) for gamma rays of shortest wave length to 10^6 m (10^8 cm) for the longest known radio waves. It should be understood that there is no line of sharp demarcation but rather a gradual transition from one type to another. For example, as the radiowaves get shorter and shorter, they merge without a break into the wave length region of longer microwaves. Similarly, the long wave length end of gamma rays is indistinguishable from the shortest X-rays.

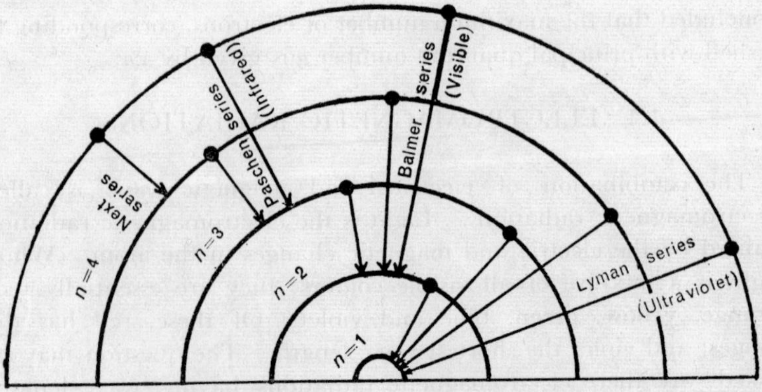

Fig. 1.8 Different electronic transitions in a hydrogen atom produce electromagnetic radiations ranging from ultraviolet to infrared.

It appears that there is no wave length that cannot, in theory, exist; the spectrum not only has no "gaps" in it, but also may be extended without limit at either end. However, we should not assume that we are able to produce or detect all possible waves, though we can now produce and measure a vast portion of the spectrum. As illustrated in Fig. 1.8, electronic transitions in a hydrogen atom produce electromagnetic radiations ranging from ultraviolet to infrared. X-rays are produced by the electronic transitions in the inner shells of the heavier atoms. Gamma rays are produced due to the changes in the nucleus. The radiowaves and the microwaves are produced by oscillations produced by suitable electronic circuits.

It is necessary to clarify here that this book is more concerned with X-rays and gamma-rays part of the electromagnetic spectrum. In fact, in the present context radiation means, X-rays, gamma-rays, electrons, protons, neutrons, alpha particles, etc. All these radiations except X-rays, are produced by the nucleus. Since X-ray emission is an atomic property, we will discuss its production and spectrum in this chapter and leave other radiations for subsequent chapters that deal with nuclear properties.

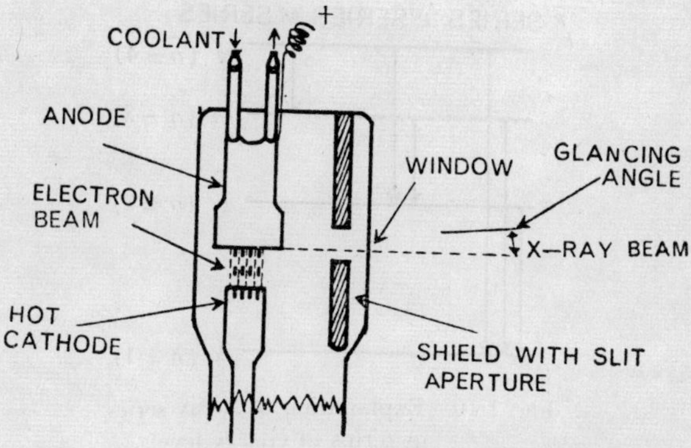

Fig. 1.9 Schematic diagram of an X-ray tube.

X-rays are generally produced by X-ray tubes. An X-ray tube is basically a big vacuum tube containing a heated cathode (electron emitter) and an anode or target as shown in Fig. 1.9. Electrons emitted by the cathode are accelerated through a high potential of many thousands of volts, to strike against a metal anode. There are two ways in which this electron bombardment may cause X-ray radiation to be emitted. First, the electrons may be rapidly decelerated when they strike the target. Since any accelerated or decelerated charge, except the electrons in atomic orbits, emits electromagnetic radiations, the deceleration of free bombarding electrons, which is not quantised produces electromagnetic radiations (photons) of all frequencies (continuous radiation) less than a certain limit. This continuous radiation is called Bremsstrahlung. The limit on frequency or wave length comes from the principle of conservation of energy; the decelerating electron cannot emit a photon whose energy is greater than the kinetic energy of the electron. Using equation (1.5) it can be shown that the shortest wave length emitted by a decelerating electron of 20 KeV energy is 6.2×10^{-9} cm. The continuous X-rays rise to a broad maximum and fall off gradually with increasing wave length. As a general rule, the fraction of kinetic energy of the electron converted into radiation in this manner increases with the energy of the electron and with the atomic number of target in which it is slowed down.

Apart from being slowed down, the bombarding electron may knock out an electron from the inner shells of an atom of the anode or target. The vacancy so created will then be immediately filled

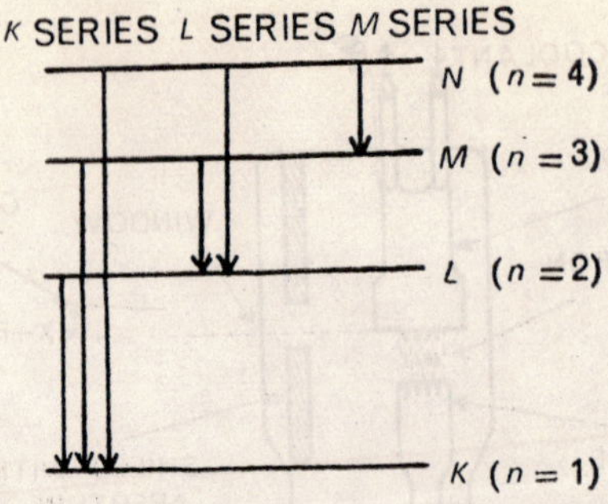

Fig. 1.10 Explanation of X-ray series in terms of energy levels.

by one of the outer electrons falling down and the energy difference between the inner and the outer electron shells is emitted as electromagnetic radiation. This is of fixed wave length and falls in X-rays region, if the electron jump takes place in the inner shells of the heavier atoms. This is because such electrons are more tightly bound to the nucleus, consequently higher energies are involved in their electronic transitions as compared to lighter atoms. It may be mentioned here that the vacancies in the atomic shell may

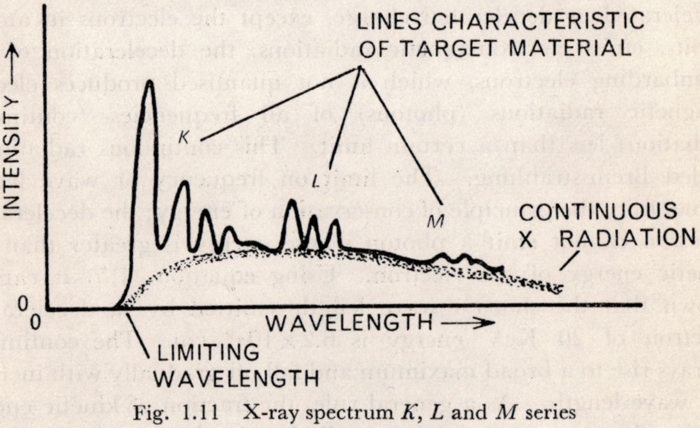

Fig. 1.11 X-ray spectrum *K*, *L* and *M* series are shown as spikes.

be created by any radiation, not necessarily by electrons whose energy is greater than the binding energy of the electrons in that shell. If the initial absorption of energy results in the ejection of one or two electrons from the *K* shell ($n=1$) level, X-rays produced are called *K* series X-rays. Similarly, X-ray lines in *L* series are obtained when the vacancy is created in the *L* shell ($n=2$) level, and so on, as shown in Fig. 1.10. The wave lengths of *K* series are smaller than *L* series. Similarly the wave lengths of *L* series are smaller than *M* series, and so on. These series, are shown as the spikes in Fig. 1.11. The series of spikes marked *K* are caused by vacancies in the innermost *K* shell being filled by electrons from the *L*, *M* and other outer shells. The spikes marked *L* are emitted by electrons from outer shells falling into *L* shell vacancies etc. The emission of these series is called X-ray fluorescence. The wave lengths for these series (*K*, *L*, *M*, etc.) vary regularly from one element to the next. This variation is used in the elemental analysis of materials by X-ray fluorescence.

CHAPTER TWO

The Nucleus

2.1 CONSTITUENTS OF THE NUCLEUS

2.1.1 Electron-Proton Model of the Nucleus

In order to explain large deflection of alpha particles by the atom, Rutherford proposed that the entire positive charge of the atom is concentrated at the centre, called the nucleus, of the atom. Can it then be assumed that the nucleus is made up of protons only, their number being equal to the number of electrons outside the nucleus to account for the electrical neutrality of the atom? The answer is no, because in that case the protons present in the atom can account for only about half the value of the atomic weight. That is, such a model cannot explain the observed atomic weights. If the number of protons are increased to account for the atomic weights, the atoms do not remain electrically neutral. In order to explain both, the observed atomic weight as well as the electrical neutrality of the atom, it was proposed that the number of protons in the nucleus are more than the number of electrons outside the nucleus and that some electrons are present in the nucleus also, such that the total number of electrons and the total number of protons are equal in the atom. Thus, it was thought that the nucleus is made up of electrons and protons. But this model is not able to explain several nuclear properties. In fact, even on size considerations, it can be said that electron cannot exist inside the nucleus because the size of the electron (10^{-8} cm) is much bigger than the nucleus (10^{-12} cm).

2.1.2 Neutron-Proton Model of the Nucleus

In 1932, J. Chadwick discovered neutron which is a neutral particle of zero charge and of mass approximately equal to that of proton. Since, electron-proton model was not able to explain several nuclear properties, it was quickly discarded after the discovery of neutron. It is now believed that the nucleus is made up of neutrons and protons. The number of protons in the nucleus is called the atomic number Z. The number of neutrons in the nucleus is called neutron number N. The sum of numbers of neutrons and protons in the nucleus is called mass number A. Thus

$$A = Z + N \tag{2.1}$$

The symbol used to denote a nuclear species is the chemical symbol of the element with the atomic number as the left subscript and the mass number as superscript; for example : ${}^{4}_{2}\mathrm{He}$, ${}^{14}\mathrm{N}$, ${}^{235}_{92}\mathrm{U}$. The atomic number is often omitted, because it is uniquely determined by the chemical symbol.

2.2 NUCLEAR TERMS

2.2.1 Nuclide

The term nuclide is used to signify any one of the known species of atoms characterised by the constitution of its nucleus, in particular by the number of protons and neutrons in its nucleus.

2.2.2 Isotopes

Isotopes are the atomic species (nuclides) of same atomic number Z, that is belonging to the same element but having different neutron number and consequently, different mass number A. For example, ${}^{12}_{6}\mathrm{C}$, ${}^{13}_{6}\mathrm{C}$, and ${}^{14}_{6}\mathrm{C}$, are isotopes of carbon because they contain the same number (six) of protons in their nucleus. Similarly, ${}^{1}_{1}\mathrm{H}$, ${}^{2}_{1}\mathrm{H}$ and ${}^{3}_{1}\mathrm{H}$ are isotopes of hydrogen (see Fig. 2.1a). ${}^{15}\mathrm{N}$ is an isotope of nitrogen and ${}^{15}\mathrm{O}$ is an isotope of oxygen and so on.

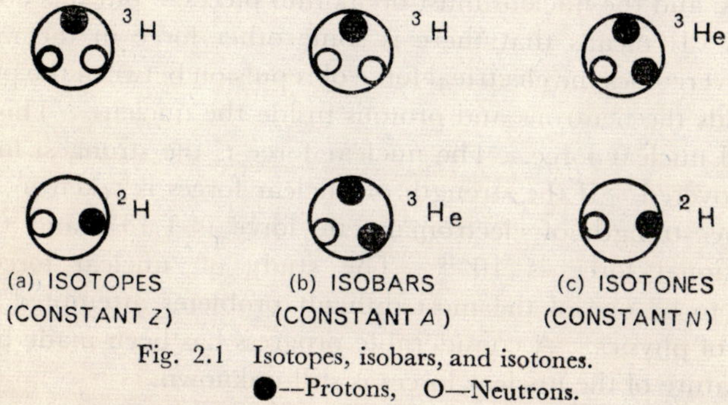

Fig. 2.1 Isotopes, isobars, and isotones.
●—Protons, O—Neutrons.

2.2.3 Isobars

Atomic species (nuclides) having the same mass number (A) but different atomic numbers (Z) are called isobars. For example, ${}^{3}_{1}H$ and ${}^{3}_{2}H$ are isobars (see Fig. 2.1b).

2.2.4 Isotones

Nuclides having the same number of neutrons (N), but different mass numbers are called isotones. For example, ${}^{3}_{1}H$ and ${}^{4}_{2}He$ are isotones because each possesses the same number (two) of neutrons (see Fig. 2.1c).

2.2.5 Isomers

Nuclides having the same atomic number and the same mass number but differing in their energy content are called isomers. That is, nuclear isomers are the different energy states of the same nucleus, each having a measurable lifetime (except the ground state which may be stable). The isomers are denoted as ${}^{A}_{Z}X$ and ${}^{Am}_{Z}X$.

2.2.6 Nucleon

Either neutron or proton is called nucleon.

2.3. NUCLEAR FORCES

The nucleus is made up of neutrons and protons. There are as many as 92 protons in the nucleus of uranium whose size is about 10^{-12} cm. The question arises as to how so many protons stay so near to each other in the nucleus. According to Coulomb's law protons must repel each other with very large force because of their nearness, and the nucleus must break into pieces. But this does not happen. It means that there is some other force in the nucleus which overcomes the electrical force of repulsion between the protons and binds the neutrons and protons inside the nucleus. This force is called nuclear force. The nuclear force is the strongest force so far discovered. If the strength of nuclear forces is taken to be 15, then the strength of electromagnetic force is 1/137, and that of gravitational force is 10^{-40}. The study of nuclear forces has proved to be one of the most difficult problems attempted in the history of physics. A considerable progress has been made but the exact nature of the nuclear forces is still unknown.

The interaction between nucleons is complex in the sense that it depends not only on the separation of the nucleons but also on their relative orientations. The interaction is of very short range, and beyond a separation of a few Fermis (1 Fermi$=10^{-13}$ cm), the nuclear force is exceedingly small. At the other extreme, the inter-nucleon force becomes strongly repulsive, probably at a separation of less than 0·5 Fermi. To compare the Coulomb and nuclear forces we find that at a separation of about 1 Fermi, the nuclear force between two protons is 30 to 40 times the strength of the Coulomb repulsion. At a separation of 4 Fermis, the nuclear and Coulomb forces are nearly equal; whereas at 25 Fermis the short range nuclear force has decreased to 10^{-6} times the electrical force. The onset of a nuclear repulsive force at very short separations prevents the collapse of nuclear matter.

In 1935, H. Yukawa suggested that the nuclear force between two nucleons may arise from the continuous exchanges of mesons between the neutrons and the protons. Yukawa predicted that the mass of this meson should be about 200 times the electron mass. Such particles called π mesons (pions) were discovered in 1947. The mass of π meson is 273 times the mass of the electron.

2.4. NUCLEAR ENERGETICS

2.4.1 Energy Mass Transformation

According to Einstein's famous theory of relativity, mass may be converted into energy and vice-versa. Their relationship is given by :

$$E=mc^2 \tag{2.2}$$

Where E is the energy, m is the mass and c is the velocity of light. Velocity of light is a very large quantity ($c=3\times 10^{10}$ cm/sec). Occurrence of c^2 in equation (2·2) means that a small amount of mass corresponds to a very large amount of energy. One gram of mass is equivalent to 9×10^{20} ergs of energy, which is produced by the explosion of 20,000 tons of TNT. For the same reason, the mass of a given amount of energy is usually very small. Thus, the heat energy which has to be supplied to 1 kg of water to raise its temperature from the freezing point to the boiling point (10^5 cal) has a mass of about 4×10^{-9} gm and the mass of light emitted per minute by a 10-watt bulb amounts only to 7×10^{-12} gm.

Another important result of the special theory of relativity is that the mass of an object is not an absolute quantity. It changes with its velocity. The mass m of an object which is moving with velocity v is given by

$$m = \frac{m_0}{1-(v^2/c^2)} \tag{2.3}$$

where m_0 is the mass of the object when it is at rest.

2.4.2 Atomic Mass Unit and Electron Volt

In nuclear science the mass is usually expressed in atomic mass unit (amu) and energy in electron volt (eV). One atomic mass unit is equal 1/12th the mass of the ^{12}C atom. The absolute value of the unit is

$$1 \text{ amu} = 1.660420 \times 10^{-24} \text{ gm} \tag{2.4}$$

One electron volt is equal to the energy gained by the electron when it is accelerated through a potential difference of 1 volt. The relationship between amu and electron volt is given by :

$$1 \text{ amu} = 931.4 \times 10^6 \text{ eV} = 931.4 \text{ MeV} \tag{2.5}$$

2.4.3 Mass Defect

The sum of the masses of the various constituents of the atom is always found to be more than the atomic weight of the atom. For example, in the case of the oxygen isotope ^{16}0, there are 8 protons, 8 neutrons and 8 electrons; the sum of their masses is equal to 16.132 amu, whereas the atomic weight of ^{16}0 is 15.99491 amu. Thus the oxygen isotope ^{16}0 is lighter than its constituents by 0.13709 amu. This difference between the total mass of the individual protons, neutrons and electrons which make up the atom and the atomic mass of the atom is called mass defect. It is given by the formula

$$\text{mass defect} = Z m_h + (A-Z) m_n - M \tag{2.6}$$

Where Z is the atomic number, m_h and m_n represent the masses of the hydrogen atom and the neutron, respectively, $A-Z$ is the number of neutrons in the nucleus and M is the atomic weight. By multiplying Z with mass of hydrogen atom (m_h) the masses of the number of protons and electrons that are present in the atom are automatically added. If, in the above equation the atomic weight is replaced by the nuclear mass, the mass of the hydrogen atom should then be replaced by the mass of proton.

2.4.4 Nuclear Binding Energy

Nuclear binding energy is the energy which is released when nuclear constituents combine to form the nucleus. The same amount of energy would be required to break up the nucleus into its constitutent particles and hence the energy equivalent to the mass defect is taken as the measure of the nuclear binding energy. Thus, if m_h m_n and M are expressed in the atomic mass units (amu), the nuclear binding energy in MeV is given by;

$$\text{B.E. in MeV} = [Zm_h + (A - Z)m_n - M]\ 931.4 \tag{2.7}$$

Binding energy per nucleon is called average binding energy. It is given by :

$$\text{B.E.}/A \tag{2.8}$$

In the previous section (2.4.3) the mass defect of the ^{16}O isotope was found to be 0.13709 amu; its binding energy therefore, would be:

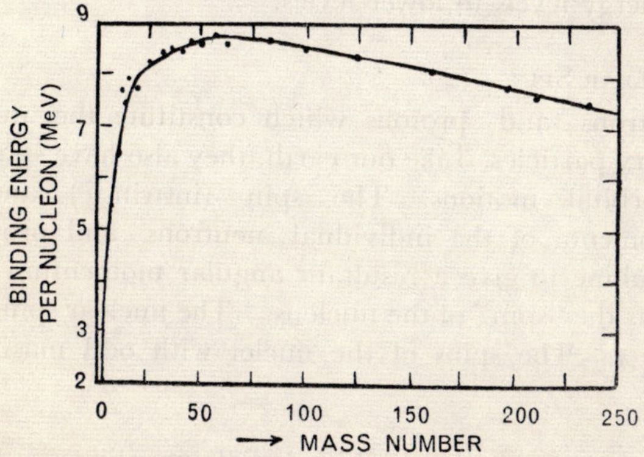

Fig. 2.2 Binding energy per nucleon as a function of mass number.

$$931.4 \times 0.13709 \text{ MeV} = 127.68 \text{ MeV}$$

There are 16 nucleons in ^{16}O. Therefore, the average binding energy in ^{16}O is $127.68/16 = 7.06$ MeV. Fig. 2.2 shows the variation of average binding energy with respect to mass number. It may be noted that the binding energy per nucleon which is relatively small for light nuclei increases with mass number, reaches a broad maximum close to 8.5 MeV in the mass number range from about 40 to 120. For higher mass numbers the values decrease and fall to 7.6 MeV for uranium. It is the diminishing of the average binding per nucleon on either side of the broad maximum which causes the release of energy in nuclear fission and fusion.

2.5 PROPERTIES OF THE NUCLEUS

2.5.1 Nuclear Energy Levels

Like atomic electrons (section 1.1.3, Fig. 1.4), the nucleus also exists in several energy states, with the difference that the energy gap between nuclear levels is much larger as compared to electronic levels. In general, the energy separation of nuclear levels is of the order of a million electron volt. If $E_2 — E_1$ in equation (1.5) is set equal to 1 MeV (or 10^6eV), the wave length of the radiation corresponding to such a transition would be about 1.2×10^{-12} cm. Such a wave length corresponds to very short X-rays or long gamma rays (see Fig. 1.7). Thus, gamma rays are emitted due to the transition of a nucleus from higher (excited) energy levels to lower levels.

2.5.2 Nuclear Spin

The neutrons and protons which constitute the nucleus are not stationary particles. Like our earth, they also have spin (instrinsic) and orbital motions. The spin (instrinsic) and orbital angular momenta of the individual neutrons and protons in a nucleus combine to give a resultant angular momentum generally referred to as the "spin" of the nucleus. The nuclear spin is usually denoted by *I*. The spins of the nuclei with odd mass numbers are:

$$I = 1/2,\ 3/2,\ 5/2,\ 7/2, \text{ etc.}$$

whereas the spins of the nuclei with even mass numbers are :

$$I = 0, 1, 2, 3, 4, \text{etc.}$$

The spin of nuclei with even mass in their ground state is usually 0 or 1.

2.5.3 Nuclear Magnetic Moment

Motion of electrical charge is called current. A wire carrying current deflects the magnetic needle, if it is brought near the wire carrying current. This means that the motion of charge generates a magnetic field. An electron (or proton) is a rotating charged particle. Therefore it behaves like a small magnetic dipole. Its magnetic moment is given by :

$$\mu = \frac{eh}{4\pi mc} \tag{2.9}$$

Where μ, is the magnetic moment, e is the electronic charge, h is Planck's constant, m is the rest mass of the electron and c is the velocity of light. The magnetic moment of an electron is called one Bohr magneton. The magnetic moment of the proton can also be calculated by equation (2.9), by replacing the mass of the electron with that of the proton. Since proton is 1836 times heavier than the electron, its magnetic moment is 1836 times less than that of electron. The magnetic moment of a proton is called one nuclear magneton. The magnetic moments of nuclei are of the order of one or a few nuclear magnetons. The magnetic properties of the electron are utilised in the working of the Electron Spin Resonance (ESR) and that of nucleus in the working of Nuclear Magnetic Resonance (NMR) spectrometers.

2.5.4 Nuclear Stability

Only certain combinations of protons and neutrons produce stable nuclei. A review of the nuclear composition of various nuclides reveals the following three main rules concerning the stability of nuclei:

I. Odd-Even Rule;

II. Magic Numbers;

III. Neutron-Proton Ratio.

I. Odd-Even Rule. Table 2.1 gives a summary of stable nuclei along with the even or odd number of protons and neutrons in them. A survey of this table brings to light some interesting points. Of the stable nuclei, the greatest number have an even number of both protons and neutrons and the least number have an odd number of both protons and neutrons. This means that even-even nuclei are the most stable and odd-odd nuclei are the least stable.

Table 2.1

Proton	Neutron	Number of stable Nuclei
even	even	164
even	odd	54
odd	even	50
odd	odd	4

II. Magic Numbers. It has been noted that a nucleus is more stable if its neutron and proton numbers are 2, 8, 20, 28, 50, 82, or 126. These numbers are called magic numbers. The nucleus is most stable if its neutron and proton numbers both are magic numbers. For example the helium nucleus (alpha-particle) with two neutrons and two protons is exceptionally stable.

III. Neutron-Proton Ratio. Fig. 2.3. gives the number of protons and neutrons for various stable nuclides. It may be noted that the neutron-proton ratio (n/p) is equal or very nearly equal to unity for light nuclides. The n/p ratio is unity for all nuclides which lie on a dotted line drawn at an angle of 45° in Fig. 2.3. Above ^{40}Ca ($Z=20$, $N=20$), no stable isotope exists with equal numbers of neutrons and protons, for as Z increases, the repulsive electrical force increases at greater rate than the attractive nuclear forces. The addition of extra neutrons is necessary to increase the average distance between protons within the nucleus to reduce the electrical repulsive force. For heavy nuclei the n/p ratio is 1.5 or greater.

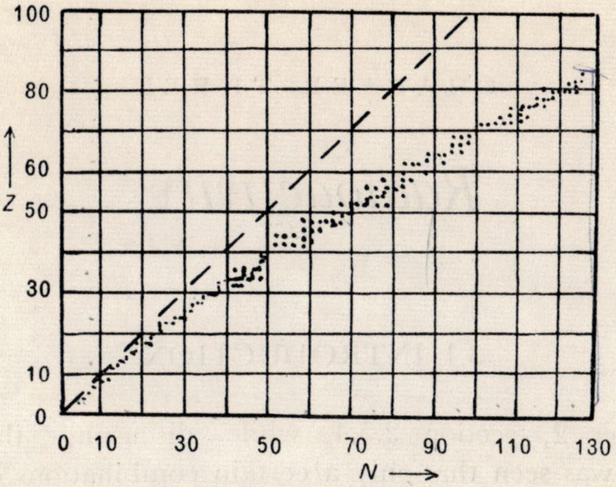

Fig. 2.3 Plot of Z versus N for stable nuclei.

If a nuclide is too rich in neutron it will be radioactive and will emit electrons as will be explained in the next chapter (section 3.3.1). Similarly if a nucleus is too rich in protons it will also be unstable and will emit a positron or capture an electron from the electronic orbits (sections 3.3.2 and 3.3.3).

CHAPTER THREE

Radioactivity

3.1 INTRODUCTION

In chapter 2, section 2.5.4, while discussing the nuclear stability, it was seen that only a certain combinations of neutrons and protons produce stable nuclei whereas others do not. Unstable nuclei decay by emitting alpha (helium nucleus), beta (electron or positron) and gamma radiations. The emission of these radiations by unstable nuclei in their decay process is called radioactivity. The decay process is called radioactive decay. The nuclei, which emit these radiations are called radioactive nuclei. If the nuclei which emit these radiations are found in nature then their decay is called natural radioactivity; but if they are produced in the laboratory, then their decay is called artificial radioactivity. Artificial radioactivity is more useful because most of the radioactive isotopes (nuclei) which are used in actual practice are artificially produced at various nuclear research centres all over the world. Radioactivity is not influenced by any chemical or physical means. That is, we can neither retard nor accelerate the rate of radioactive decay by use of heat, pressure, electrical fields, etc.

Not all the three radiations (alpha, beta and gamma) are emitted by the same nucleus in radioactivity. Usually, a radioactive nucleus emits alpha-particles and gamma rays or beta-particles and gamma-rays. That is, if a radioactive source emits alpha-particles, it also usually emits gamma rays. The same thing is true for beta-active sources also. Of course, there are several exceptions to this rule, like ^{3}H, ^{14}C, ^{32}P, etc. These nuclei emit only beta rays and are called pure beta emitters.

3.2 ALPHA-DECAY

The alpha-particle is made up of two neutrons and two

protons. Its charge number is 2 and mass number is 4. Therefore, if a nucleus emits alpha-particles, its charge number is reduced by 2 and mass number by 4 because the charge number and the mass number are conserved in the nuclear transformations. Alpha-decay can be expressed by the following generalised equation:

$$^{A}_{Z}X \rightarrow ^{A-4}_{Z-2}Y + ^{4}_{2}He + Q \tag{3.1}$$

where Q is the amount of energy released in the alpha-decay (an example of conversion of mass into energy). This energy is shared between alpha-particle and the residual nucleus in accordance with the laws of conservation of energy and momentum. Equation (3.2) represents the alpha-decay of a specific nucleus, radon-222:

$$^{222}_{86}Rn \rightarrow ^{218}_{84}Po + ^{4}_{2}He + 5.58MeV \tag{3.2}$$

Although, the total energy released in the α-decay of radon-222 is 5.58 MeV, the energy of the α-particle is only 5.48 MeV. The difference of 0.10 MeV is the energy of recoil imparted to the newly formed polonium nucleus to conserve the momentum.

Alpha-particles emitted in a particular decay are monoenergetic. In the decay of thorium-228 for example, at least five specific values of alpha particle energy are possible, one corresponding to each decay pathway as shown in Fig. 3.1. It may be noted that while 71% of the thorium-228 nuclei decay directly to the ground state of radium-224 by emission of a 5.42 MeV alpha-particle, 28% of the nuclei emit a 5.34 MeV alpha-particle in decaying to an excited state of radium-224 which is 0.084 MeV above the ground state. This excited state ultimately decays to the ground state by emitting gamma-ray of 0.084 MeV energy. Other excited states also decay to ground state by emitting γ-rays as shown in Fig. 3.1.

It has been found with few exceptions that nuclei which emit alpha particles have the atomic number of 83 or more. This is not unexpected, because alpha-particle is relatively a large particle. Therefore, a nucleus which emits such a particle must itself be large. Further, there exists an inverse relation between alpha-particle energy and half-life of its emitter. That is, an alpha active nucleus with shorter half-life emits alpha-particles of higher energy and vice-versa.

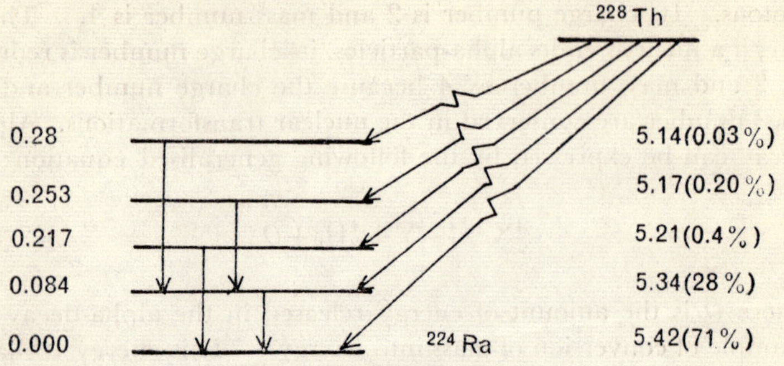

Fig. 3.1 Decay scheme of ^{228}Th.

3.3 BETA-DECAY

The following three processes are called beta-decay:

(1) Emission of electron (β^-) by the nucleus.

(2) Emission of positron (β^+), positively charged electron, by the nucleus.

(3) Electron capture (EC) by the nucleus.

Electrons and positrons do not constitute the nucleus. How then, are they emitted by the nucleus? They are created in the nucleus only at the time of their emission. When a nucleus has an excess of neutrons, that is, if its neutron-proton ratio is more than the value for the stable nucleus in a particular region (see Fig. 2.3), then it will be unstable because of the deficiency of protons. Such a nucleus will achieve stability by changing its neutron into proton. It is represented in the following way:

$${}_0n \rightarrow {}_1p + {}_{-1}e + {}_0\bar{\nu} \text{ (antineutrino)} \quad (3.3)$$

Antineutrino is a particle with zero charge and practically no mass. It is very difficult to detect this particle because its interaction with matter is negligible.

When a nucleus has more protons than permitted by the neutron-proton ratio for stable nuclei in certain region, it will be unstable. Such a nucleus will achieve stability by converting its protons into neutrons. This conversion can be accomplished by either positron (β^+) emission

$${}_1p \rightarrow {}_0n + {}_{+1}e + {}_0\nu \quad \text{(neutrino)} \quad (3.4)$$

or by electron capture (EC)

$$_{1}p+_{-1}e \text{ (orbital electron) } \rightarrow {}_{0}n+{}_{0}\nu \tag{3.5}$$

3.3.1 β-Decay

A nucleus emits β^- if its neutron-proton ratio is too big. The β-decay process has been represented by equation (3.3). It can also be written in the form:

$$n \rightarrow p+\beta^-+\bar{\nu} \tag{3.6}$$

A free neutron will decay with a half-life of about 12 minutes. But the situation is considerably different when the neutron is bound within the nucleus. The half-life of the β-emitting nuclei is not governed by the half-life of free neutron. In β-decay the charge of the nucleus increases by one but its mass number remains unaltered.

β^- decay of phosphorus-32 is illustrated by:

$$^{32}_{15}\text{P} \rightarrow {}^{32}_{16}\text{S}+\beta^-+\nu+1.71 \text{ MeV} \tag{3.7}$$

That is, after emitting β^-, the ^{32}P nucleus changes to sulphur-32. β-decay can be represented by the generalised equation:

$$^{A}_{Z}\text{X} \rightarrow {}^{A}_{Z+1}\text{Y}+\beta^-+\bar{\nu}+Q \tag{3.8}$$

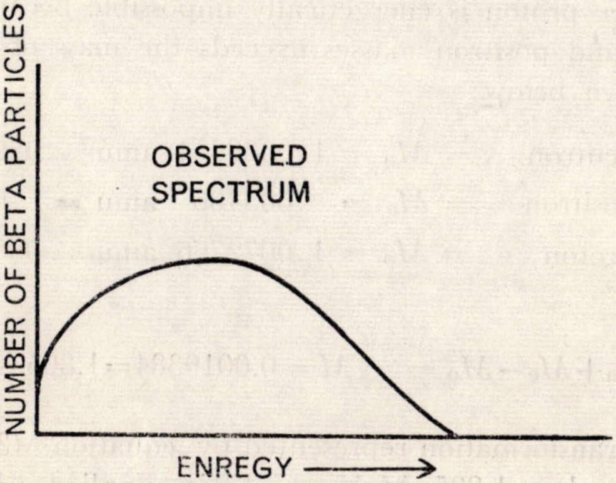

Fig. 3.2 Beta spectrum.

The energy Q released in the process is shared between the electron and the antineutrino; energy taken by the recoiling residual nucleus being practically zero. That is why unlike alpha-particles which are monoenergetic from a given source, beta-particles are emitted with a range of energies lying between zero and a maximum energy E_m for a particular isotope as shown in Fig. 3.2. The antineutrino, corresponding to the electron of zero energy, takes away the total energy released in β^--decay. Similarly the antineutrino corresponding to the maximum energy E_m of the electron gets no energy. This sharing of energy between electron and antineutrino produces continuous spectrum. Maximum energy, E_m of β^- particles is the characteristic of a specific isotope, which emits these particles. That is why only E_m for a particular isotope is given in the literature. For example when we say that ^{32}P emits 1.71 MeV electrons, it only means that the maximum energy of β^- emitted by ^{32}P is 1.71 MeV. Actually, electrons produced in the decay of ^{32}P have energies ranging from zero to 1.71 MeV.

3.3.2 β^+-Decay

β^+-decay occurs if the neutron-proton ratio is not enough for the stability of the nucleus. That is, those nuclei emit positrons which are deficient in neutrons. This process has been represented by equation (3.4). It can also be written in the form:

$$p \rightarrow n + \beta^+ + \nu \qquad (3.9)$$

Decay of free proton is energetically impossible because the sum of neutron and positron masses exceeds the mass of the proton. This is shown below:

Mass of neutron $= M_n = 1.0086654$ amu $= 939.550$ MeV

Mass of positron $= M_e = .005486$ amu $= 0.511$ MeV

Mass of proton $= M_p = 1.0072766$ amu $= 938.256$ MeV

Thus,

$$M_n + M_e - M_p = \triangle M = 0.0019384 = 1.805 \text{ MeV}.$$

Therefore, transformation represented by equation (3.9) cannot take place unless 1.805 MeV energy is supplied. Free proton cannot get this energy. But the proton contained within the nucleus may get this energy, provided the atomic mass of the decaying nucleus is greater than the residual nucleus by at least $2m_0c^2 = 1.022$MeV (m_0 is the rest mass of the electron). A minimum of 1.022MeV energy is required to create a positron (creation of positron is an example of the conversion of energy into mass). Thus unlike free neutrons free protons are stable. Conversion of proton into neutron, positron and the neutrino is possible only when the proton is contained in the nucleus.

In β^+-decay the mass number remains constant, while the charge number decreases by 1. Positron decay of zinc-65 is represented as :

$$^{65}_{30}\text{Zn} \rightarrow \ ^{65}_{29}\text{Cu} + \beta^+ + \nu \qquad (3.10)$$

Positron decay can be represented by the generalised equation:

$$^{A}_{Z}\text{X} \rightarrow \ ^{A}_{Z-1}\text{Y} + \beta^+ + \nu + Q \qquad (3.11)$$

The energy Q released in the β^+ decay is shared between β^+ and ν. Therefore, similar to β^-, the spectrum of β^+ is also continuous.

Every positron emitted in β^+ decay is annihilated on coming into contact with an electron, because positron is the antiparticle of electron. Therefore, positron will have a very short life in our environment so rich in electrons. The amount of energy released in the annihilation of a positron is equal to the energy (1.022 MeV), which is spent in its creation. This energy is generally released as two gamma rays each of 0.511 MeV energy. They are emitted in opposite directions as required by the law of conservation of momentum. Only in very rare cases three gamma-rays (photons) each of 0.341 MeV are emitted in the same plain at angles of 120° to each other.

3.3.3 Electron Capture (EC)

A nucleus which is deficient in neutrons decays either by β^+-decay or electron capture. In electron capture a nucleus captures one of its own orbital electrons and emits a neutrino. Because a K-electron is captured in about 90% of the events, this process is often called K-capture. The electron capture can be represented by

$$p + e^- \rightarrow n + \nu \qquad (3.12)$$

In electron capture decay the charge number of the decaying nucleus is reduced by one, its mass number remains the same. Electron capture is expressed by the generalised equation:

$$ {}^{A}_{Z}X + {}^{0}_{-1}e \rightarrow {}^{A}_{Z-1}Y + {}^{0}_{0}\nu \qquad (3.13) $$

Electron capture and β^+ decay both result in reducing the charge number of decaying nucleus by one. Both these processes (EC and β^+-decay) usually occur in the same nucleus (see Fig. 3.9). In some cases where sufficient energy (1.022 MeV) is not available for the creation of β^+, electron capture is the only process of decay (see Fig. 3.6).

In the electron capture, the vacancy is filled by an electron from higher energy levels. The process is accompanied by the emission of X-rays characteristic of the residual (daughter) nucleus. For example, the decay of ^{55}Fe by EC results in the emission of the characteristic X-ray of manganese. It is by means of X-rays that a nuclide decaying by EC can be detected. Auger effect competes with the emission of X-rays in EC. Auger effect is the de-excitation of an atom by electron emission. The electron emitted in Auger effect is called Auger electron.

3.4 GAMMA-DECAY

Decay of nuclides either by alpha or by beta-decay processes results in the formation of daughter nuclides more often in their excited states (see Figs. 3.1, 3.3, 3.11). These excited states usually decay instantaneously by gamma emission as shown in Fig. 3.3. Since both the mass number and the charge number of gamma-ray are zero, in gamma-decay the mass number and the charge number of the de-exiciting nucleus does not change.

3.5 INTERNAL CONVERSION (IC)

Excited nuclides may de-excite by the emission of orbital electrons through a coupling between the excited nucleus and the orbital electron. This process is known as internal conversion (IC). The ejected electron is called conversoin electron. The electron is usually ejected from *K*-shell. Its energy *I* is given by :

$$ T = E - B \qquad (3.14) $$

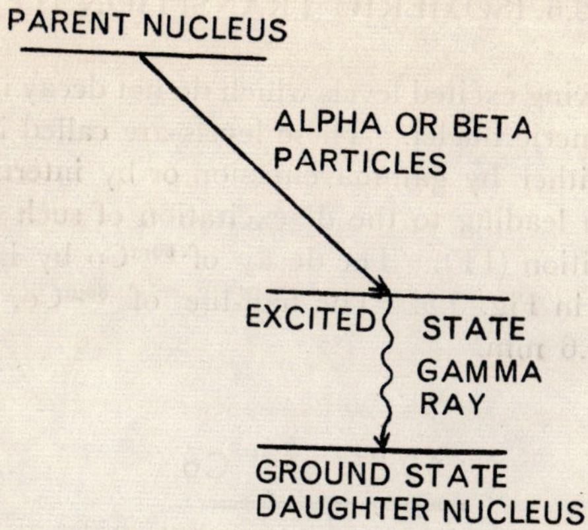

Fig. 3.3 Emission of gamma radiation in radioactive decay.

where E is the energy of the excited state and B is the binding energy of the emitted electron. Conversion electrons therefore have certain discrete energy values and appear as peaks in β spectra, as shown in Fig. 3.4.

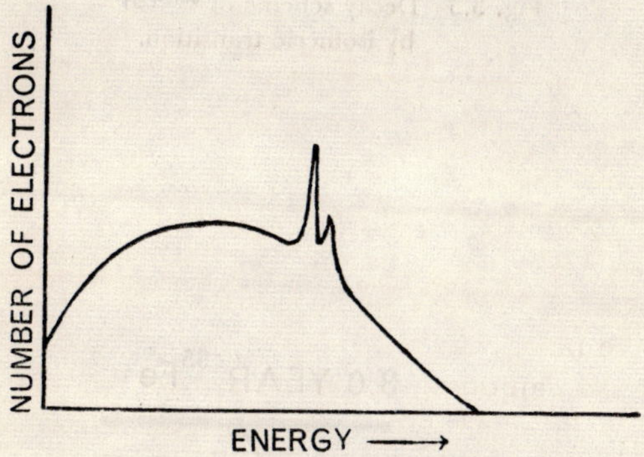

Fig. 3.4 Two groups of conversion electrons are superimposed on beta spectrum.

The internal conversion coefficient (denoted by α) is the ratio of the number of conversion electrons ($\mathcal{N}_e$) to the number of gamma photons ($\mathcal{N}_r$) emitted per unit time in the de-excitation of a given state:

$$\alpha = \frac{\mathcal{N}_e}{\mathcal{N}_r} \tag{3.15}$$

3.6 ISOMERIC TRANSITION (IT)

Nuclides having excited levels which do not decay instantaneously are called isomeric nuclei. These levels are called isomeric levels. They decay either by gamma emission or by internal conversion. The transition leading to the de-excitation of such levels is called isomeric transition (IT). The decay of ^{60m}Co by isomeric transition is shown in Fig. 3.5. The half-life of ^{60m}Co, isomeric level of ^{60}Co, is 10.6 min.

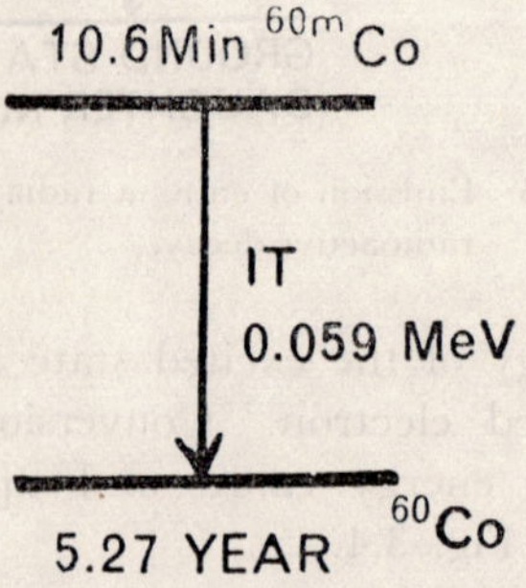

Fig. 3.5 Decay scheme of ^{60m}Co by isomeric transition.

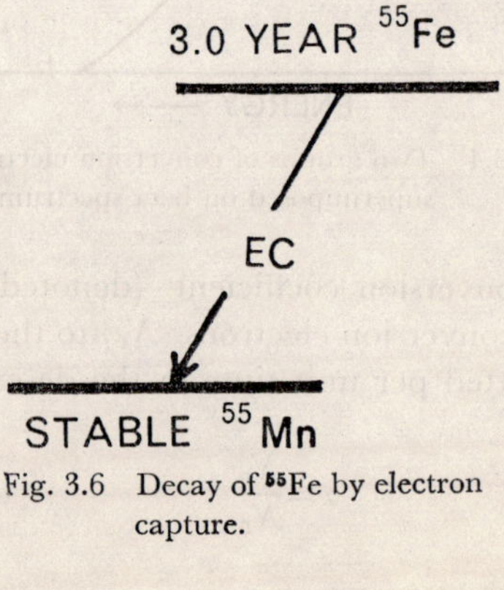

Fig. 3.6 Decay of ^{55}Fe by electron capture.

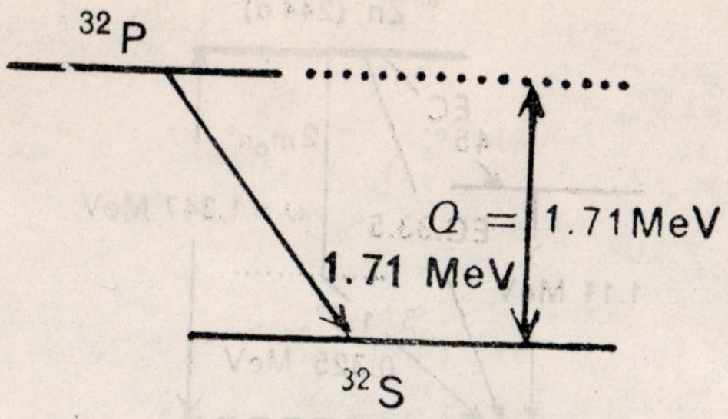

Fig. 3.7 Decay scheme of ^{32}P (14d).

3.7 DECAY SCHEMES

Radioactive decay may occur by any one, or by a combination of several processes that are described in the previous sections of this chapter. The decay scheme of a nucleus represents in quantitative way the decay processes (modes) of the nucleus. The decay schemes of nuclei which decay by single mode are simple. For example, ^{55}Fe decays to ^{55}Mn only by electron capture; its simple decay scheme is shown in Fig. 3.6. The decay schemes of many pure beta emitters (nuclei which do not emit gamma-ray following beta-decay) are simple. Commonly used pure beta emitters are ^{3}H, ^{32}P, ^{14}C, ^{35}S, ^{45}Ca, ^{90}Sr. Their decay schemes are similar. The decay scheme of ^{32}P belonging to pure beta emitter class is shown in Fig. 3.7.

The decay schemes of nuclei which decay by a combination of several processes are complex. This is true for the decay schemes of ^{198}Au (see Fig. 3.8) and ^{65}Zn (see Fig. 3.9). The decay scheme of ^{131}I shown in Fig. 3.10 is an example of an especially complex decay scheme. In the decay of ^{131}I beta-particles of four

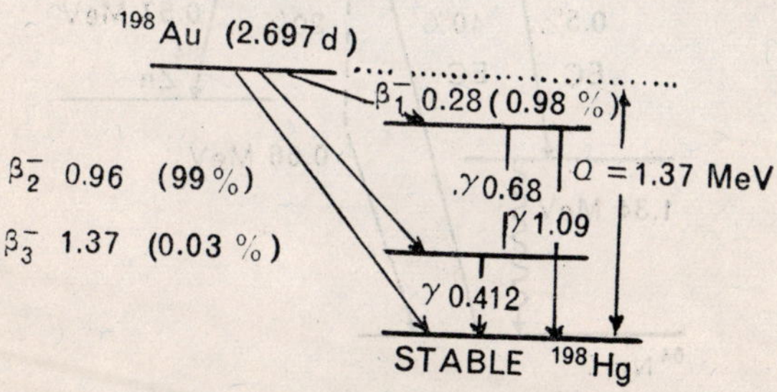

Fig. 3.8 Decay scheme of ^{198}Au.

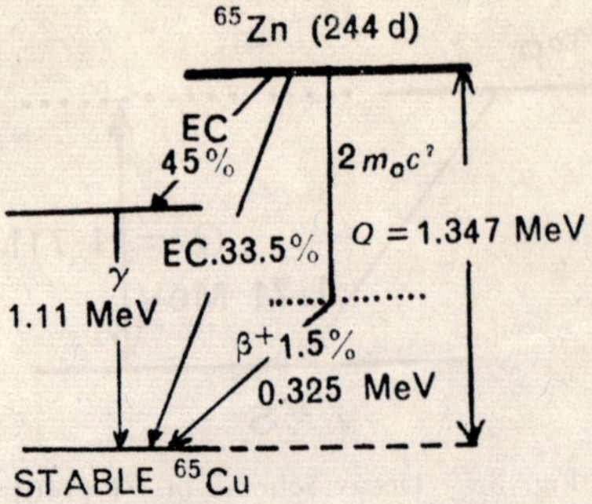

Fig. 3.9 Decay scheme of ^{65}Zn. The vertical line labelled $2m_0c^2$ represents the amount of energy required to create an electron pair.

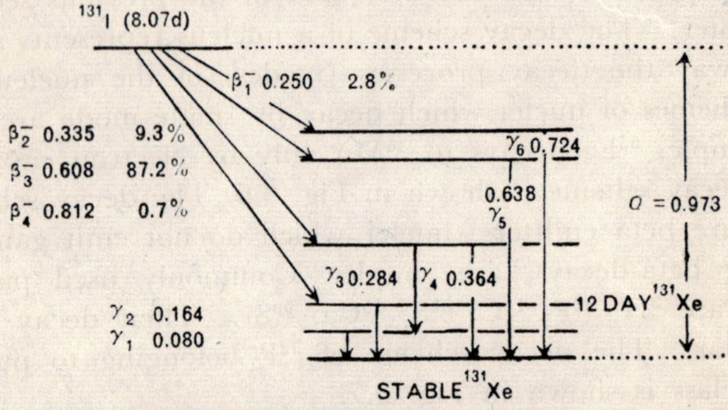

Fig. 3.10 Decay Scheme of ^{131}I.

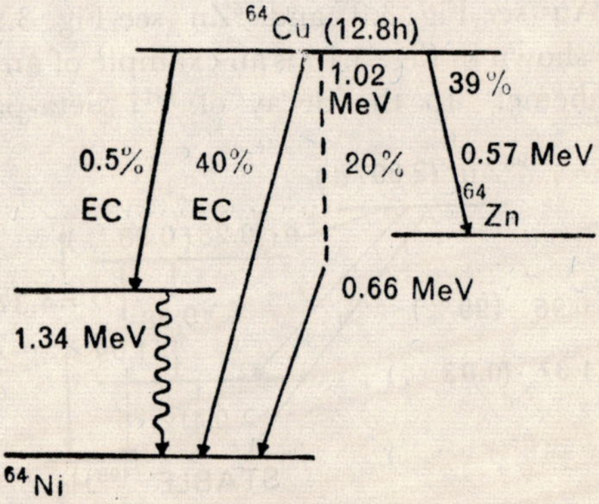

Fig. 3.11 Decay scheme of ^{64}Cu.

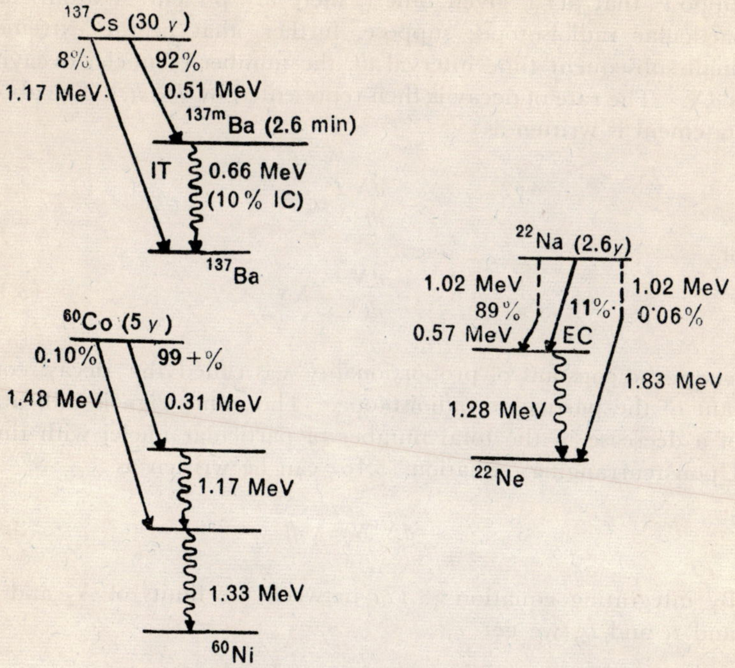

Fig. 3.12 Decay schemes of ^{137}Cs, ^{60}Co and ^{22}Na.

distinct energy transitions are emitted. In 87.2% transitions the disintegration results in the emission of beta particles with maximum energy E_m=0.608MeV. Six gamma-rays are emitted in the decay of ^{131}I which are shown by vertical lines, the most prominent of them has the energy of 0.364MeV. Fig. 3.11 shows the decay scheme of ^{64}Cu which decays by all the three (β^+, EC and β^-) beta-decay modes. The decay schemes of ^{137}Cs, ^{60}Co, and ^{22}Na, which are often used for the calibration of gamma-ray spectrometers are shown in Fig. 3.12.

3.8 EQUATIONS OF RADIOACTIVE DECAY AND GROWTH

3.8.1 Radioactive Decay Rate

The rate of radioactive decay of nuclei at any time is proportional to the total number of radioactive nuclei present at that time. In order to express this statement in mathematical form, suppose that at a given time t, there are present N atoms of a particular radioisotope; suppose, further, that in the extremely small subsequent time interval dt, the number of nuclei decaying is dN. The rate of decay is then represented by dN/dt. The above statement is written as

$$-\frac{dN}{dt} \propto N$$

or

$$-\frac{dN}{dt}=\lambda N \tag{3.16}$$

where the constant of proportionality λ is called the decay constant of the particular radioisotope. The minus sign is indicative of a decrease in the total number of particular nuclei with time. Upon rearranging, equation (3.16) can be written as

$$-dN/N=\lambda\ dt \tag{3.17}$$

By integrating equation (3.17) between the limits of N_1 and N_2 and t_1 and t_2, we get

$$\int_{N_1}^{N_2} \frac{dN}{N} = -\int_{t_1}^{t_2} \lambda\ dt \tag{3.18}$$

or $$\ln\ (N_2/N_1) = -\lambda\ (t_2-t_1) \tag{3.19}$$

where N_1 is the number of radioactive nuclei at time t_1, and N_2 is their number present at time t_2. In equation (3.19) if we put $t_1=0$ and $N_1=N_0$ and $t_2=t$ and $N_2=N$, we will get

$$\ln\ N/N_0=-\lambda t \tag{3.20}$$

where N_0=number of a particular radioactive nuclei present initially;
N= number of that nuclei remaining at time t.

Expressing in exponential form, we get :

$$N=N_0e^{-\lambda t} \tag{3.21}$$

where e is the base of natural logarithms.

Rate of disintegration $\left(\frac{dN}{dt}\right)$ is called absolute A. The activity A at any time t is given by:

$$A = -\frac{dN}{dt} = \lambda N \tag{3.22}$$

substituting the value of N from equation (3.21) into equation (3.22) we get :

$$A = N_0 \, e^{-\lambda t} \tag{3.23}$$

The initial activity A_0 is:

$$A_0 = \lambda N_0 \tag{3.24}$$

substituting this value of N_0 in equation (3.23), we get

$$A_0 = Ae^{-\lambda t} \tag{3.25}$$

This is the expression for the absolute activity A at time t in terms of the initial absolute activity A_0 and the decay constant λ. In actual practice we usually measure relative activity. Equation (3.25) is valid for the relative activity also.

3.8.2 Half-life (T) and its Determination

The half-life of a radioactive isotope is equal to the time required to reduce its activity to half the initial value. It is also defined as the time required to reduce the active nuclei to half the initial value. Let the activity of sample is A_1 at time t_1, and A_2 at time t_2.

Then

$$A_1 = A_0 e^{-\lambda t_1} \tag{3.26}$$

and

$$A_2 = A_0 e^{-\lambda t_2} \tag{3.27}$$

Dividing equation (3.26) by (3.27)

$$\frac{A_1}{A_2} = e^{\lambda (t_2 - t_1)} \tag{3.28}$$

If $t_2 - t_1 =$ half-life (T);

then according to the definition of the half-life,

$$A_2 = A_1/2 \tag{3.29}$$

substituting equation (3.29) in (3.28), we get

$$2 = e^{\lambda(t_2 - t_1)} = e^{\lambda T} \tag{3.30}$$

$$\text{In } 2 = \lambda T \tag{3.31}$$

$$T = \frac{2.3 \log 2}{\lambda} \tag{3.32}$$

$$T = \frac{0.693}{\lambda} \tag{3.33}$$

If the half-life is known, the decay constant can be calculated and vice-versa.

If the activity A of radioactive sample with single decay constant, is plotted against time t, a curve as shown in Fig. 3.13 will be obtained.

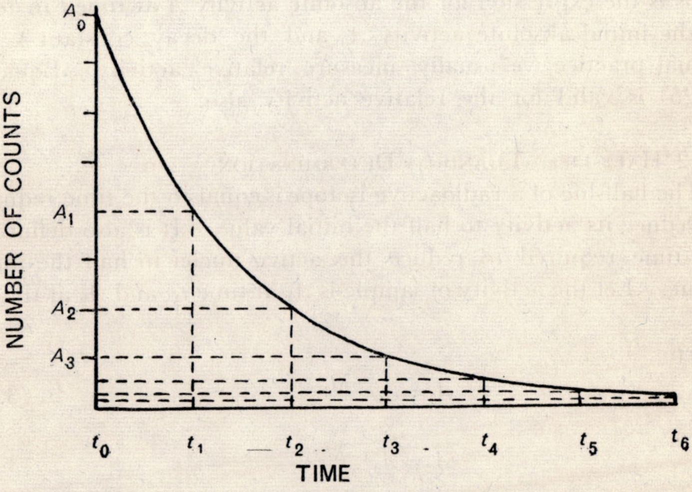

Fig. 3.13 Radioactive decay curve on linear scale.

It may be noticed that the original activity A_0 reduces to $A_1 = A_0/2$, in time t_1. Therefore t_1 is the half-life of this particular radioisotope.

Similarly, A_1 is reduced to $A_2=A_1/2$ in time (t_2-t_1).

Obviously:

$$t_1=t_2-t_1=t_3-t_2=t_4-t_3=t_5-t_4=t_6-t_5=\text{half-life} \qquad (3.34)$$

Thus, the plot presented in Fig. 3.13 can be used for determining the half-life. However, in practice such plot is not made, instead a straight line plot as shown in Fig. 3.14 is made which can be drawn more accurately than the curve line plot of Fig. 3.13. The equation for the straight line is obtained by taking the natural logaritham of equation (3.25) :

$$\text{In } (A/A_0)=-\lambda t \qquad (3.35)$$

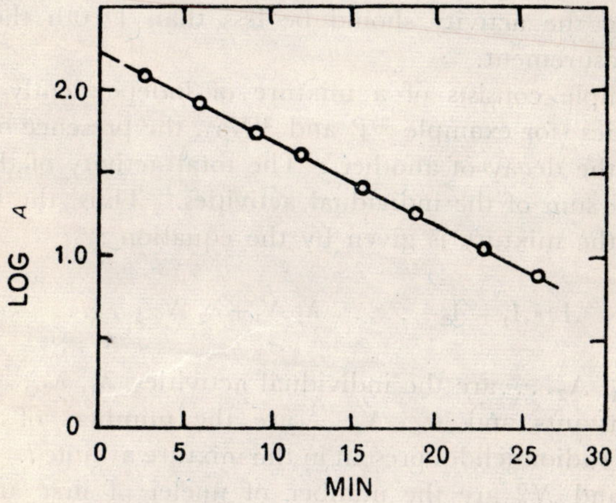

Fig. 3.14 Radioactive decay curve on semi-log scale.

In (A/A_0) means the logarithm of (A/A_0) to the base e. The conversion to ordinary logarithm to the base 10 may be made by introducing the conversion factor 0.4343, so that

$$\log (A/A_0)=-0.4343\ \lambda t \qquad (3.36)$$

which may be written as

$$\log A=\log A_0-0.4343\ \lambda t \qquad (3.37)$$

This equation is similar to the familiar equation of straight line ($y=mx+c$). Here:

$$\left.\begin{aligned} y&=\log A \\ m&=-0.4343\lambda \\ x&=t \\ c&=\log A_0 \end{aligned}\right\} \tag{3.38}$$

The equation (3.37) means that if the logarithm of activity (A) is plotted against the time, the result would be a straight line as shown in Fig. 3.14. The slope of this line will be equal to $-0.4343\ \lambda$. By knowing λ, half-life can be calculated from equation (3.33). An easier way to determine the half-life is to read directly the time in which the activity is reduced to half, as explained in connection with Fig. 3.13. An important care that should be taken in preparing the straight line plot is that the period of counting the activity should be less than 1/10th the half-life under measurement.

If a sample consists of a mixture of independently decaying radionuclides (for example ^{32}P and ^{24}Na), the presence of one will not effect the decay of another. The total activity of the sample will be the sum of the individual activities. Thus, the total activity A of the mixture is given by the equation :

$$A=A_1+A_2+\ldots.=\lambda_1 N_1+\lambda_2 N_2+\ldots. \tag{3.39}$$

Where, A_1, A_2....are the individual activities, λ_1, λ_2,.... are the decay constants and N_1, N_2....are the numbers of atoms of individual radionuclides present in the mixture at time t.

If N_1^0 and N_2^0 are the number of nuclei of first and second type respectively, at zero time, then according the equation (3.25)

$$N_1=N_1^0\ e^{-\lambda_1 t}$$

and

$$N_2=N_2^0\ e^{-\lambda_2 t}$$

substituting these values of N_1 and N_2 in the equation (3.39), we get

$$A=N_1^0\ e^{-\lambda_1 t}+N_2^0\ e^{-\lambda_2 t}+\ldots. \tag{3.40}$$

If two radioisotopes are mixed together, they will still decay as they did separately but the log of the sum of the two activities plotted against time will not be a straight line as shown in Fig. 3.14, but a curved line as shown in Fig. 3.15. If the two activities have widely different half-lives, the curve is concave upward. This curvature is produced because the shorter-lived component becomes less significant as time passes. In fact, after sufficient time, the longer-lived activity will entirely predominate and its half-life may be read from the later protion of the decay curve. In order to determine the half-life of shorter-lived activity, the last portion of the decay curve (Fig. 3.15) which is a straight line is extrapolated back to $t=0$ and the extrapolated line is subtracted from the original curve. The resultant curve represents the decay of shorter-lived activity and is a straight line. The half-life can be easily read from the straight line decay curve. In principle this kind of analysis can be applied to samples consisting of many activities with widely different half-lives. But in practice experimental uncertainties in the observed data may be expected to make it difficult to handle systems of more than three components.

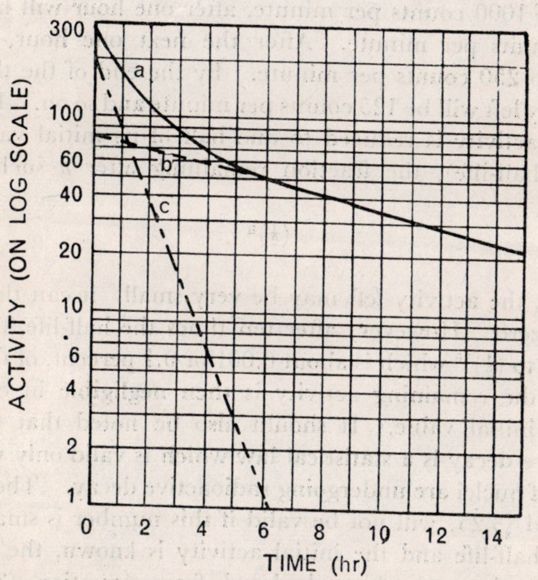

Fig. 3.15 Composite decay curve of two mixed radioisotopes and its analysis into individual components.
(a) Composite decay curve.
(b) Longer-lived component ($t_{\frac{1}{2}}=8.0$ hr.)
(c) Shorter-lived component ($t_{\frac{1}{2}}=0.8$ hr.)

The method of measuring the half-life by the observation of direct decay of the radioisotopes as explained above can be used if the half-life ranges from a few seconds to about 10 years. For measuring half-lives in the range of 10 to 10^{18} years the 'weighing method' is used. From the equation (3.22)

$$A \text{ (activity)} = N\lambda.$$

The decay constant λ, and hence the half-life can be determined, if the number of atoms N in the source and its absolute activity A are known. N can be calculated by weighing and using atomic weight and Avogadro's number. Absolute activity A can be determined by counting the radiation with the detector of known efficiency (see section 5.13). For the measurement of a very short half-life concidence technique is used.

The variation in activity of a radioactive isotope with respect to time is exponential in nature. It means that if a particular radioactive isotope has a half-life of one hour, then an initial activity of 1000 counts per minute, after one hour will be reduced to 500 counts per minute. After the next one hour, it will be reduced to 250 counts per minute. By the end of the third hour, the activity left will be 125 counts per minute and so on. In general, since the activity is reduced to one half of its initial value in the time T (half-life), the fraction remaining after n such intervals will be

$$\left(\tfrac{1}{2}\right)^n \qquad (3.41)$$

Although, the activity left may be very small, it can theoretically never be zero. However, after ten times the half-life, the activity has fallen to $\left(\frac{1}{2}\right)^{10}$ which is about 0.001 or 0.1 percent, of the original amount; the remaining activity is then negligible in comparison with the initial value. It should also be noted that the law of radioactive decay is a statistical law which is valid only when large number of nuclei are undergoing radioactive decay. The equations (3.21) and (3.25) will not be valid if this number is small.

If the half-life and the initial activity is known, the activity at any other time can be calculated from equation (3.25). For example, if initially the activity of ^{32}P, whose half-life is 14.3 days is 1000 counts per minute, then after 10 days, its activity would be:

$$A = A_0\, e^{-\lambda t} = 1000\; e^{-0 \cdot 693(10)/14 \cdot 3} = 606 \text{ cpm}$$

3.8.3 Average Life (τ)

The average life expectancy of a particular radioactive nuclei is found by the sum of the times of existence of all the nuclei divided by their initial number. If N is very large, this sum may be approximated to an equivalent integral, thus

$$\tau = -\frac{1}{N_0}\int_{t=0}^{t=\infty} t\, dN = \frac{1}{N_0}\int_0^{\infty} t\lambda N\, dt$$

$$\tau = \lambda\int_0^{\infty} t\, e^{-\lambda t}\, dt = -\left[\frac{\lambda t+1}{\lambda}\, e^{-\lambda t}\right]$$

$$\tau = \frac{1}{\lambda} \tag{3.42}$$

It may be noted by comparing equations (3.33) and (3.42) that the average life is greater than the half-life by a factor of 1.443. The average life can be easily calculated if the half-life is known.

3.8.4 Successive Radioactive Decay

The decay of a radioactive nucleus may result in the formation of a daughter nucleus which is also radioactive. The radioactive daughter in turn decays thus becoming parent to produce a new daughter nucleus which may be stable or radioactive . This type of series decay can be expressed as:

$$A \xrightarrow{\lambda_1} B \xrightarrow{\lambda_2} C$$

where A decays to B and B decays to C which is stable, and λ_1, and λ_2 are the decay constants of A and B respectively.

Let N_1 and N_2 represent the number of atoms of A and B, respectively present at time t, and N_1^0 and N_2^0 be the number of A and B respectively at zero time, then

$$N_1 = N_1^0\, e^{-\lambda_1 t} \tag{3.43}$$

$$\frac{dN_1}{dt} = -\lambda_1 N_1 \tag{3.44}$$

$$\frac{dN_2}{dt} = \lambda_1 N_1 - \lambda_2 N_2 \tag{3.45}$$

Substituting equation (3.43) into (3.45), we get

$$\frac{dN_2}{dt} + \lambda_2 N_2 - \lambda_1 N_1^0 e^{-\lambda_1 t} = 0 \tag{3.46}$$

The solution of this linear differential equation is

$$N_2 = \frac{\lambda_1}{\lambda_2 - \lambda_1} N_1^0 \left(e^{-\lambda_1 t} - e^{-\lambda_2 t}\right) + N_2^0 e^{-\lambda_2 t} \tag{3.47}$$

In the special case where only nuclei of type *A* are present initially, i.e., $N_2^0 = 0$ at $t = 0$,

$$N_2 = \frac{\lambda_1}{\lambda_2 - \lambda_1} N_1^0 \left(e^{-\lambda_1 t} - e^{-\lambda_2 t}\right) \tag{3.48}$$

Secular Equilibrium. If the decrease in the activity of parent is insignificant during many half-lives of the daughter, i.e., if the half-life of the parent is many times greater than the daughter or $\lambda_2 >>> \lambda_1$, then equation (3.48) reduces to

$$N_2 = \frac{\lambda_1}{\lambda_2} N_1^0 e^{-\lambda_1 t} \tag{3.49}$$

and substitution of equation (3.43) into (3.49) gives

$$N_1 \lambda_1 = N_2 \lambda_2 \tag{3.50}$$

That is, in secular equilibrium the activity of daughter will be equal to that of parent; and the daughter will decay with the half life of the parent. These conclusions are illustrated in Fig. 3.16. For the occurrence of secular equilibrium the difference between the half-lives of parent and daughter is usually a factor of 10^4 or more. The decay of radium-226 with the half-life of 1620 years

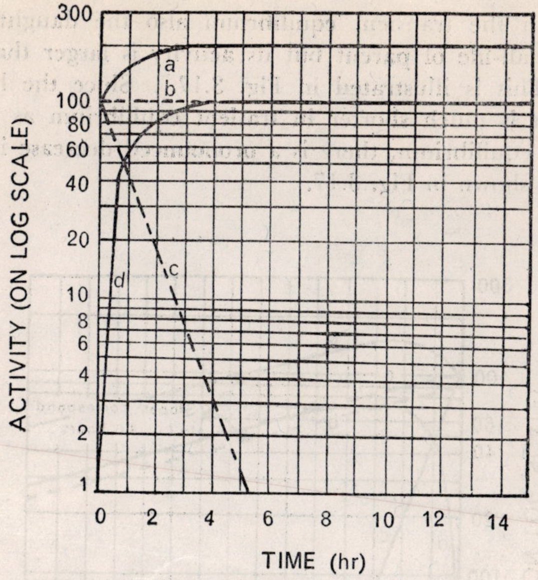

Fig. 3.16 Secular equilibrium.

(a) Total activity of an initially pure parent fraction.

(b) Activity due to parent ($t_{\frac{1}{2}} = \infty$). This is also the total daughter activity in parent-plus-daughter fractions.

(c) Decay of freshly isolated daughter fraction ($t_{\frac{1}{2}} = 0.8$ hr.)

(d) Daughter activity growing in freshly purified parent fraction.

to radon-222 which has the half-life of only 3.8 days is a typical example in which secular equilibrium is attained.

Transient Equilibrium. If the half-life of the parent is greater than that of the daughter by a small factor (about 10) rather than a large factor (10^4 or more), that is, $\lambda_2 > \lambda_1$, then equation (3.46) takes the form:

$$N_2 = \frac{\lambda_1}{\lambda_2 - \lambda_1} N_1^0 e^{-\lambda_1 t} \tag{3.51}$$

because as t becomes very large, $e^{-\lambda_2 t}$ becomes negligible compared to $e^{-\lambda_1 t}$. Since $N_1 = N_1^0 e^{-\lambda_1 t}$, we get:

$$N_1 \lambda_1 = N_2 \lambda_2 - N_2 \lambda_1 \tag{3.52}$$

That is, in the transient equilibrium also the daughter decays with the half-life of parent but its activity is larger than that of parent. This is illustrated in Fig. 3.17. Since the half-life of the parent is much shorter in trasient equilibrium as compared to secular equilibrium, there is a pronounced decrease in its total activity as shown in Fig. 3.17.

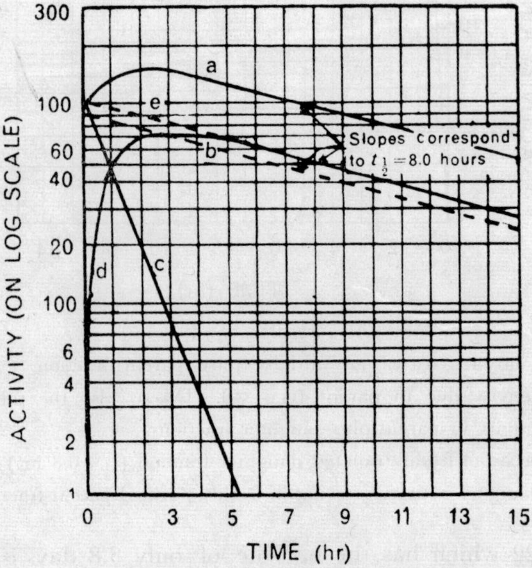

Fig. 3.17 Transient equilibrium.

(a) Total activity of an initially pure parent fraction.

(b) Activity due to parent ($t_{\frac{1}{2}}$=8.0 hr.)

(c) Decay of freshly isolated daughter fraction ($t_{\frac{1}{2}}$=0.80 hr).

(d) Daughter activity growing in freshly purified parent fraction.

(e) Total daughter activity in parent-plus-daughter fractions.

No Equilibrium. If the parent is shorter-lived than the daughter, i.e., $\lambda_1 > \lambda_2$, no equilibrium is attained at any time. If one starts with pure parent, then as the parent decays, the amount of daughter will rise, pass through a maximum and finally decay with the characteristic half-life of the daughter. This is illustrated in Fig. 3.18.

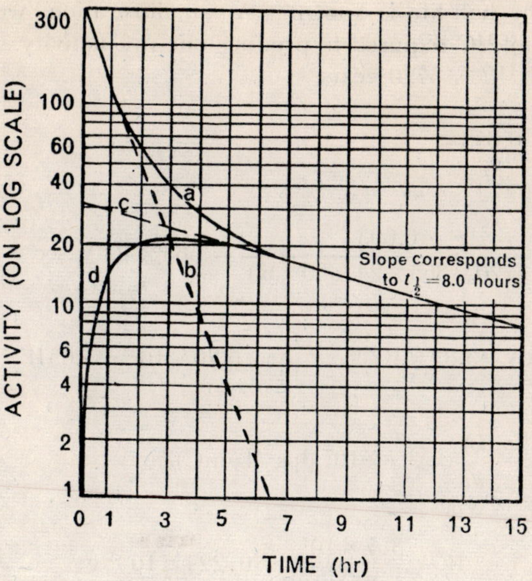

Fig. 3.18 No equilibrium.

(a) Total activity.
(b) Activity due to parent ($t_{\frac{1}{2}}$=0.80 hr).
(c) Extrapolation of final decay curve to time zero.
(d) Daughter activity in initially pure parent.

3.9 UNITS OF RADIOACTIVITY

Curie, which is denoted by C, is the unit of radioactivity. It expresses the rate of disintegration of a radioactive source. A source is said to possess an activity of one curie (1 C) if 3.7×10^{10} of its atoms disintegrate per second. The disintegrations per second is abbreviated as dps. In practice, the activities are expressed in terms of megacuries, curies, millicuries and microcuries.

1 megacurie	(MC)	$= 3.7 \times 10^{16}$ dps $= 10^{6}$ curies
1 curie	(C)	$= 3.7 \times 10^{10}$ dps
1 millicurie	(mC)	$= 3.7 \times 10^{7}$ dps $= 10^{-3}$ curie
1 microourie	(μC)	$= 3.7 \times 10^{4}$ dps $= 10^{-6}$ curie.

The mass *W* of any radioactive atom required to produce a given amount of activity can be easily calculated from equation

(3.22) if the half-life is known. As an illustration, we calculate the weight of ^{14}C reqiured to produce one mC activity. The half-life (T) of ^{14}C is 5720 years.

$$\lambda = \frac{0.693}{T}$$

$$= \frac{0.693}{5720 \times 365 \times 24 \times 60 \times 60} = 3.83 \times 10^{-12} \text{ sec}^{-1}$$

$$-\frac{dN}{dt} = \lambda N = 3.83 \times 10^{-12} \times \frac{W}{14} \times 6.20 \times 10^{23} = 1.65 W \times 10^{11} \text{ sec}^{-1}$$

Putting $-\dfrac{dN}{dt} = 3.7 \times 10^{7}$ dps (1 mC)

$$W = \frac{3.7 \times 10^{7}}{1.65 \times 10^{11}} = 0.224 \times 10^{-3} \text{ g}$$

Thus 0.224 mg of ^{14}C will produce an activity of 1 mC. In practice, however, a radioisotope will be accompanied by a variable quantity of stable isotopes of the same element. This stable form is called *carrier.* The concentration of radioisotope in an element or compound is specified by the term *specific activity* which is defined as activity per unit mass of the sample. By some procedures radioactive samples can be prepared which do not contain carrier. Such samples are called *carrier-free* (abbreviated C.F.). For example, in a solution of carrier-free sodium-radiophosphate, there would exist no stable phosphorus.

CHAPTER FOUR

Interaction of Nuclear Radiations with Matter

4.1. INTRODUCTION

Nuclear radiations (alpha, beta, proton, neutron, gamma-ray, etc.) are emitted as a result of various transformations and adjustments that take place inside the nucleus. Any radiation is detected by its interaction with matter. If this interaction is very small as in the case of neutrino, the detection of radiation becomes extremely difficult. It is necessary to study the manner in which nuclear radiations interact with matter in order to understand the methods and instruments used for the detection, measurement and characterisation of nuclear radiations. The harmful effects of nuclear radiations on tissues are highly dependent on the ability of the radiation to ionise the matter. Further, the selection of proper shielding material for the safe handling of radioactive substances is based on our knowledge of the penetration of nuclear radiation in matter. For example, alpha-particles emitted in radioactivity can be stopped by a sheet of paper, beta-particles may penetrate hundred such sheets, whereas gamma-rays can penetrate very large thickness of such paper. Neutrons are more penetrating than alpha, beta, and gamma-rays. Nuclear radiations can be classified in the following four groups.

1. Heavy charged particles (alpha, proton, deuteron etc.)
2. Light charged particles (mainly electrons)
3. Neutral particles (mainly neutrons)
4. Electromagnetic radiations (gamma-rays)

The passage through matter of these radiations will be treated separately because they each involve different characteristic processes.

4.2 INTERACTION OF HEAVY CHARGED PARTICLES WITH MATTER

Alpha-particles may be taken as the representative of this class of particles. When an alpha particle moves through matter, it ionises or excites the atom to which it comes sufficiently close. These changes are caused by the electrical (Coulomb) force, which can give sharp impulse to the electrons that are situated near the path of the moving charged particles.

The energy transferred to the electrons represents a loss of kinetic energy of the moving particle, which therefore will slow down and eventually stop. The loss of kinetic energy in a nuclear encounter would be much larger, but such collisions are extremely rare compared to the interaction with atomic electrons, because the area of cross-section of the nucleus is about 10^8 times less than that of the atom, and the probability of interaction is roughly proportional to the area of cross section. Hence nuclear collisions do not contribute appreciably to the overall energy loss. For kinetic energy larger than about M_0c^2 where M_0 is the rest mass of the particle, energy loss by emission of electromagnetic radiation (Bremsstrahlung) becomes increasingly important.

A crude idea of the important concepts of the energy loss in excitation and ionisation of atoms can be obtained by assuming that the charged particle is heavy and that it collides with a free electron. This assumption is valid as long as the velocity of the heavy charged particle is much greater than that of the electron or $(m/M)E >> I$, where m and M are the masses of electron and heavy charged particle respectively, E is the energy of the heavy charged particle and I is approximately the ionisation potential of the electron. Based on these assumptions the expression for the rate of loss of energy (dE/dx) which is also called the stopping power of the medium is given by

$$\frac{dE}{dx} = \frac{4\,Z^2e^4n}{mv^2}\left[\ln\frac{2mv^2}{I} - \ln\,(1-\beta^2) - \beta^2\right] \qquad (4.1)$$

Where Z is the charge number of the moving particle,

e is the charge of the electron,

n is the number of electrons per unit volume of the absorber through which the heavy charged particle is moving,

m is the mass of electron,

v is the velocity of heavy charged particle,
I is the effective ionisation potential of the atoms in absorber,
$\beta = v/c$ where c is the velocity of light.

If the velocity of the particle, v, is very small compared to c, i.e., for $\beta << 1$, then equation (4.1) reduces to

$$-\frac{dE}{dx} = \frac{4\,Z^2 e^4 n}{mv^2}\ \text{In}\ \frac{2mv^2}{I} \tag{4.2}$$

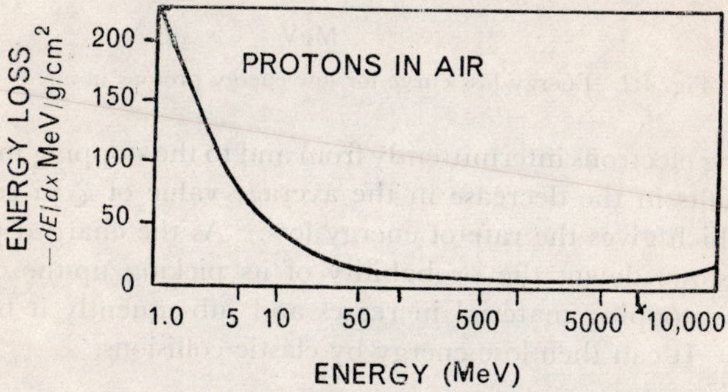

Fig. 4.1 Rate of loss of energy of proton in air.

Equation (4.2) predicts that the rate of loss of energy $(-dE/dx)$ will continuously decrease with increase in velocity (energy). On the other hand if $\beta \rightarrow 1$, the term $(1-\beta^2)$ in equation (4.1) is the most rapidly varying one and the rate of loss of energy increases with increasing energy. Thus the rate of loss of energy goes through a broad minimum which occurs at approximately twice the rest energy of the charged particle as shown in Fig. 4.1. Other important feature of equation (4.1) is that the rate of loss of energy in a given absorber of all the heavy charged particles moving with the same velocity is proportional to the square of their charges. That is, the rate of loss of energy of protons of energy E, deuterons of energy $2E$ and tritium of energy $3E$ are all the same in a given absorber and is equal to one quarter the rate of loss of energy of ^{3}He of energy $3E$ and of alpha-particle of energy $4E$.

At very low energies, (less than 0.3MeV protons) the rate of energy loss of the charged particles decreases rather than increases with the decrease in its energy as shown in Fig. 4.2. This is because, at low energies, the heavy charged particle starts picking

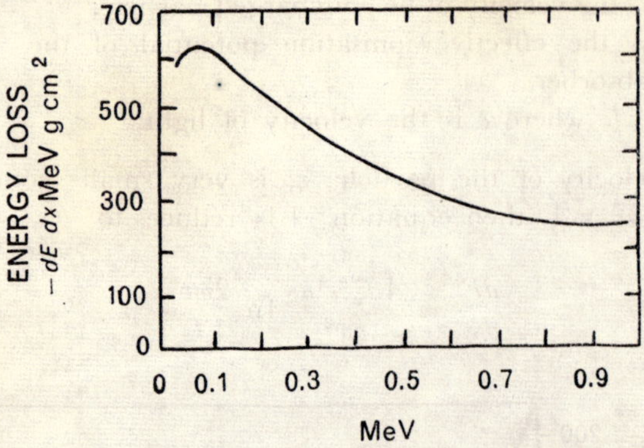

Fig. 4.2 Energy loss curve for low energy protons in air.

and losing electrons intermittently from and to the stopping material. This results in the decrease in the average value of Z in equation (4.2), which gives the rate of energy loss. As the charged particle further slows down, the probability of its picking up the electron from the stopping material increases and subsequently it becomes neutral. It can then lose energy by elastic collisions.

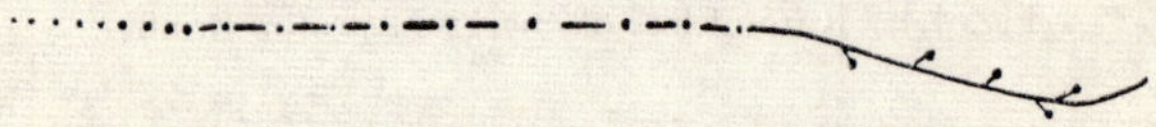

Fig. 4.3 Track of a heavy charged particle.

The distance travelled by a particle before losing its energy completely is called the range of the particle. A heavy charged particle loses a very small fraction of its energy in a single collision with an electron and it does not appreciably deflect in this collision because of its very high mass as compared to electron. Because of these reasons the path of such a particle is nearly a straight line. A typical path of a heavy charged particle is shown in Fig. 4.3 and is called the track of the particle. The thickness of the track represents the rate of energy loss; higher the rate of energy loss thicker is the track. The thinnest end of the track gives the direction through which the particle enters the stopping material. As expected, the track is thicker towards the end of the range, but again becomes slightly thinner at the end of the range. Small branchings from the main track are called delta-rays. They are produced by the secondary electrons which receive enough energy from the heavy charged particle to produce their own ionisation.

Identical charged particles, all having the same initial energy do not all have exactly the same range. Fig. 4.4 shows the number of alpha-particles found in a gas at a distance r from the source, plotted against r; the source emits alpha-particles of single energy. It is seen that the ranges of all particles of same energy in a given medium are not exactly the same but show a small spread. This phenomenon is called straggling and is caused by the statistical fluctuations in the number of collisions and in the energy loss per collision. The distance R obtained by dropping the perpendicular from the point of inflection on the abscissa as shown in Fig. 4.4 is called the mean range.

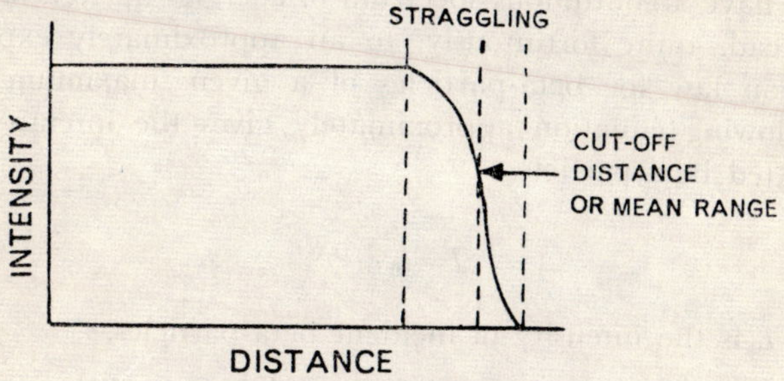

Fig. 4.4 Rate at which alpha-particles can be detected at various distances from the source.

The mean range of an alpha-particle is related to its energy in the following way :

$$E=2.12R^{2/3} \tag{4.3}$$

Where E is the energy of the alpha-particle in MeV, and R is its range in air measured in centimeters. Alpha-particles of about 1–10 MeV energies can travel several centimeters in air and several microns (10^{-3} mm) in aluminium. As the entire energy of an alpha-particle is lost in a relatively thin layer of absorber, its rate of energy loss (specific ionisation) is very high.

4.3 INTERACTION OF ELECTRONS WITH MATTER

Electrons lose energy by causing excitations and ionisations in the stopping material just as do heavy charged particles. But since the mass of electron is very small, it is scattered more and penetrate relatively deeper into matter and has a lower specific ionisation. Because of the small mass, the bremsstrahlung is a very important way by which high energy electrons lose energy. The ratio of energy loss by bremsstrahlung to energy loss by ionisation in an element of atomic number Z is approximately equal to $EZ/800$ where E is the energy in MeV. Thus in heavy materials like lead the radiation loss (bremsstrahlung) becomes appreciable even at 1 MeV, whereas in light materials (air, aluminium) it is not so important, at least for the energies available from beta emitters.

The above processes of interaction of electrons with matter coupled with the fact that the beta-particles emitted from radioisotopes have a continuous spectrum of energies up to a maximum of E_m lead, quite fortuitously, to an approximately exponential absorption law for beta-particles of a given maximum energy. The following equation approximately gives the intensity of the transmitted beta particles:

$$I = I_0 \, e^{-\mu x} \tag{4.4}$$

where I_0 is the intensity of incident beta-particles,

I is the intensity of transmitted beta-particles,

μ is called the beta-absorption coefficient of the absorber,

x is the thickness of the absorber.

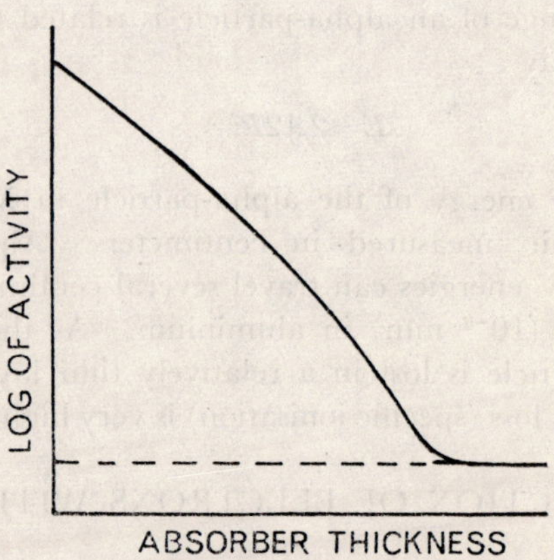

Fig. 4.5 Typical beta-particle absorption curve.

Equation (4.4) is similar to radioactive decay equation (3.25). Therefore, when the transmitted intensity is plotted as a function of the absorber thickness on semi-log paper, nearly a straight line is obtained over a portion of curve as shown in Fig. 4.5. The curve becomes practically horizontal at "R" the "range" of beta-particle. Although, all the beta-rays are stopped at this absorber thickness, one still finds some transmission of radiation. This is because of the non-characteristic X-rays (bremsstrahlung) produced by the beta-particles in the stopping materials. There are empirical relations between the maximum energy of beta-particles and their range. One of them which is valid above 0.6 MeV is

$$R=0.543E-0.160 \tag{4.5}$$

where R is the range in grams per square centimeter and E is the maximum energy in MeV.

Because of the small mass the back scattering of beta-particles is very significant. The back scattered intensity increases with increasing thickness of a back scattering material. For thickness greater than about one-third the range of the electrons, saturation is achieved, and further increase in thickness does not add to the back scattered intensity. The ratio of measured activity of a beta source with back scattering material to the activity without the back scatterer is known as back-scattering factor. The saturation back-scattering factor is essentially independent of the maximum beta-particle energies above about 0.6 MeV, and varies from about 1.3 for aluminium to about 1.8 for lead. These factors are dependent on the particular counting arrangement used, and should be determined for each configuration. In the measurement of beta activity, absorption and scattering in the sample cover, detector cover, and intervening air are very significant. Side scattering into the detector from the counter shield is also important. These effects will influence the counting rate. If the sample is not infinitely thin, self-scattering and self-absorption will all take place in the sample itself and this will cause an overall reduction in the measured activity, which is large for thick samples and small for thin samples. The counting rate at first increases by increasing the thickness of sample, then becomes constant because the contribution from the lower layers of the sample is entirely absorbed in the upper ones.

Positron loses energy more or less in the same way as the electron.

However, when the kinetic energy of the positron becomes very low, it is annihilated giving rise usually to two photons as discussed in Chapter 3, Section 3.3.2.

4.4 INTERACTION OF NEUTRONS WITH MATTER

Because neutron is electrically neutral, it cannot lose enery by ionisation. Nuclear encounters, though rare are the only possible means of its energy loss. Most of these collisions are elastic, that is, the struck nucleus is not excited, but in some cases the inelastic excitation also contributes to the energy loss. Possible interaction processes are:

(1) Scattering

(2) Nuclear Reactions

(3) Fission

(4) Free decay.

4.4.1 Scattering

It is always possible for a neutron to be scattered without exciting the struck nucleus. Such a scattering is called elastic scattering. Most neutrons lose their energy by this type of scattering. The speed of the neutrons after head-on elastic collision is given by :

$$V = V_0(M-m)/(M+m) \qquad (4.6)$$

where V is the speed of the neutron after head-on elastic collision, V_0 is its velocity before the collision, M is the mass of the struck nucleus, m is the mass of the neutron. It may be noted from the above equation that the speed of the neutron after the head-on collision would be reduced nearly to zero if the mass of the struck nucleus is equal to the mass of neutron. Since the proton and neutron masses are nearly equal, the whole of the neutron energy is transferred to the proton in a head-on collision. That is why light elements, especially hydrogen, have the best neutron slowing down (moderating) qualities. The energy transferred to the hydrogen nuclei (protons) causes ionisation by which fast neutrons are detected.

4.4.2 Nuclear Reactions

The Coulomb barrier of the nucleus does not inhibit the entry of neutrons into a nucleus, because it is electrically neutral. On an average, capture of a neutron by a nucleus is accompanied by the release of 7-8 MeV energy. The 'compound nucleus' produced by neutron capture has the same charge but the mass number is increased by one unit; it is, therefore, the adjacent isotope to the target (capturing) nucleus. This reaction is represented as:

$$^{A}_{Z}X + n \rightarrow {}^{A+1}_{Z}X + \gamma \tag{4.7}$$

and is called (n, γ) reaction. This is the most probable reaction with slow neutrons. Owing to the phenomenon of resonance, the reaction probability rises sharply in the region of certain neutron energies, which are characteristics of each isotope.

If the excitation energy of the 'compound nucleus' is sufficiently large, charged particles such as protons, deuterons, tritons or alpha-particles may be emitted. Since the Coulomb barrier must be overcome by the outgoing charged particle, the emission of the charged particle occurs in reaction between fast neutrons and light or medium-weight nuclei. There are a few reactions in which charged particles are ejected even by slow neutrons like,

$$^{10}B + n \text{ (slow)} \rightarrow (^{11}B) \rightarrow {}^{7}Li + \alpha + \gamma \tag{4.8}$$

The slow neutrons are detected by the ionisation caused by the alpha-particles produced in the above reaction.

4.4.3 Fission

In this process the compound nucleus, formed after the absorption of neutron, divides into two fragments of comparable mass, and usually a few neutrons are also ejected. The phenomenon is almost entirely limited to the heaviest nuclei. Since the fragments are charged and possess considerable energy, of the order of 100 MeV the process is useful in neutron detection. The following reactions with uranium are typical,

$$^{235}U + n \text{ (slow or fast)} \rightarrow A + B$$

$$^{238}U + n \text{ (fast)} \rightarrow C + D$$

A, B, C and D are strongly ionising fission fragments.

4.4.4 Free Decay

The free neutron decays to proton, electron and antineutrino with a half-life of 12 minutes.

4.5 INTERACTION OF GAMMA-RADIATION WITH MATTER

For a collimated beam of monoenergetic gamma-rays (photons) the intensity transmitted through a certain thickness of the absorber is given by:

$$I = I_0 e^{-\mu x} \tag{4.9}$$

Where I_0 is the initial intensity of collimated monoenergetic photons,

I is the transmitted intensity,

x is the thickness of the absorber,

μ is the absorption coefficient of the absorber for the concerned energy.

This equation is similar to radioactive decay equation (3.25). The thickness at which I_0 is reduced to half its intensity is called 'half-thickness' (analogous to half-life). If the half-thickness is expressed in g/cm^2, its value is a function of the energy of gamma-rays and is nearly independent of the type of stopping material for 0.5 to 5 MeV photons.

There are three main processes by which gamma-rays interact with matter:

(1) Photoelectric effect

(2) Compton scattering

(3) Pair production

4.5.1 Photoelectric Effect

Photoelectric effect is manifested when a gamma quantum, penetrating matter with total absorption of its energy, ejects an electron from one of the inner shells of an atom.

In most of the cases the ejected electron will come from the K-shell and this electron thereby attains a kinetic energy of :

$$E = h\nu - I_b \tag{4.10}$$

or, in other words, a kinetic energy (E) equal to the difference between the energy ($h\nu$) of the absorbed gamma quantum and the binding energy (I_b) of the ejected electron. For this process to take place, it is essential that the energy of the gamma quantum is larger than the binding energy of the ejected electron.

Upon recovery of the normal configuration of the shell or shells, the binding energy I_b is emitted as a single X-ray quanta or as Auger electron, or both. This takes place after a very short space of time.

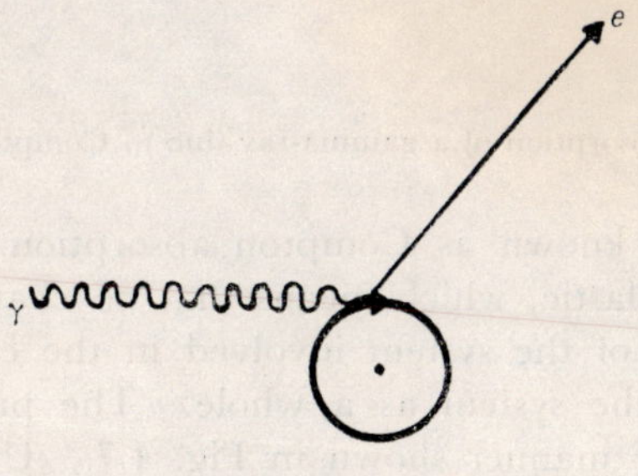

Fig. 4.6 Absorption of a gamma-ray due to photoelectric effect.

The empty place is usually filled by one of the more remote shells and very seldom by free electrons. The energy of the X-ray quantum is then $I_{bL}-I_{bK}$, if the hole in the K-shell is filled from the L-shell ; if the hole is filled from the M-shell the energy is $I_{bM}-I_{bK}$ and so on. The photoelectric effect is depicted in Fig. 4.6. The X-rays produced are absorbed within the material around the point where they occur. In this way the total energy (E_0) of the original gamma quantum is converted into kinetic energy of electrons. The probability of absorption depends not only on the energy of the original photon, but also on the atomic number of the particular absorber. When gamma rays are absorbed by aluminium, photoelectric absorption plays the greater part in energies of $\leqslant$100 KeV ; in the case of lead the figure is $\leqslant$1MeV.

4.5.2 Compton Absorption or Scattering

The process whereby a gamma quantum on collision with free electrons (for electrons which in relation to the gamma energy are only loosely bound) imparts only a part of its energy (E_c) to

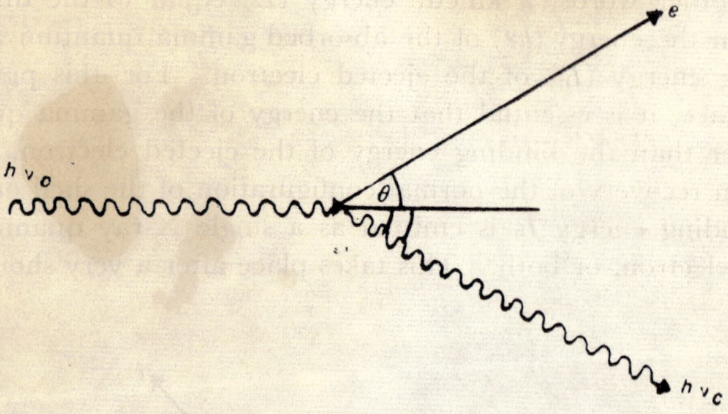

Fig. 4.7 Absorption of a gamma-ray due to Compton scattering.

the electron, is known as Compton absorption. Such collisions are said to be elastic, which means that no change occurs in the internal energy of the system involved in the collision or in the momentum of the system as a whole. The process can be represented in the manner shown in Fig. 4.7. Upon collision, the gamma-ray behaves in the same way as a particle, that is, it glances off at an angle and proceeds with reduced energy ($E_0 - E_c$). The energy of the gamma photon after the collision depends upon the angle θ at which it is deflected ; this can be expressed as :

$$h\nu_c = \frac{m_0c^2}{1 - \cos\theta + \dfrac{m_0c^2}{h\nu_0}} \tag{4.11}$$

where $h\nu_c$=energy of the photon after collision

m_0=rest mass of the electron

c=speed of light

$h\nu_c$=energy of the photon before collision.

The colliding gamma photon transfers the greatest quantity of its energy when it rebounds directly back in its track ; the maximum energy which can be imparted to the electron is

$$E_{max} = \frac{h\nu_0}{1 + \dfrac{m_0c^2}{2h\nu_0}} \tag{4.12}$$

The energy transferred can thus amount to anything from zero to this maximum.

With increasing energy of the incident ray a preference occurs for wider angles θ, which implies a preference for the transfer of larger amounts of energy to the Compton electron. In the Compton process it is not necessary to take into account the binding energy of the electron involved in the collision, but the energy of the gamma-ray must be high compared with the binding energy of the electron for Compton absorption to take place.

Further absorption of the gamma quantum with reduced energy may occur either as a result of photoelectric effect or of subsequent Compton absorption. Here again, the probability of Compton absorption depends on the energy of the incident quantum as well as on the atomic number of the absorber. By way of illustration it may be said that Compton absorption plays the major part in absorption by aluminium at energies of 100 keV to 1.5 MeV, as against 500 keV to 5 MeV for lead.

More or less the same conditions apply to the mechanism of Compton absorption as to the photoelectric effect, namely: the chances of it decrease in proportion to the energy of the incident gamma quantum. This is less marked in Compton scattering, so that, above a certain energy value (dependent on the absorbing matter) the likelihood of Compton interaction taking place is greater than the chances of photoelectric effect.

4.5.3 Pair Production

Absorption by pair production is a process that assumes im-

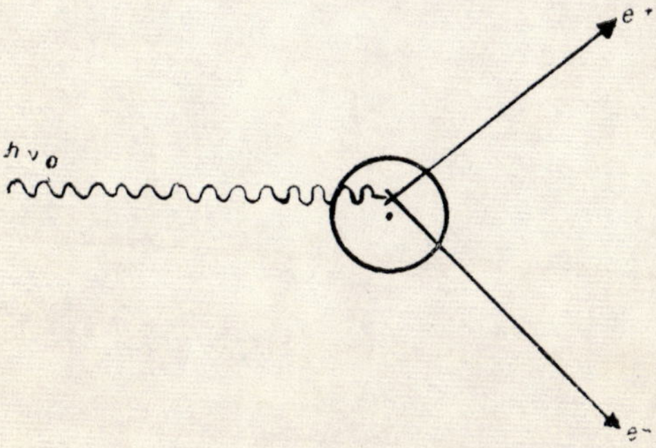

Fig. 4.8 Absorption of a gamma-ray due to pair production.

portance only in the case of gamma-rays of energy greater than 1.02 MeV, and where the absorber is heavy matter such as lead. The process is illustrated diagramatically in Fig. 4.8. It can be regarded as being the reverse of positron annihilation process (Chapter 3, Section 3.32). In the production of pairs the gamma quantum is wholly absorbed and is converted into a combination of an electron and a positron; the remaining energy after the production of electron and positron is randomly distributed between the two particles. The electron loses energy by ionisation and excitation. The position also excites and ionises the media. It soon gets annihilated producing generally two gamma-rays, each of energy 0.51 MeV.

CHAPTER FIVE

Nuclear Radiation Detection and Measurement

5.1 INTRODUCTION

Nuclear radiation detectors can be divided into the following two main groups:

(1) Electrical pulse detectors,

(2) Track detectors.

In the first type of detectors, a nuclear particle is detected by the electrical pulse generated by the particle in the detector. In the second type, the track (path) of the particle is recorded. The second type of detectors can give more information about the nuclear radiation (particle) because one can see the actual path of the nuclear particle and can have a permanent record of this path (by photography) which can be analysed at any later time. In fact looking at such photographs, one gets the feeling of the presence of such particles. They are usually employed to detect rare nuclear events and it is in this field that they have superiority over the first type of detectors. But the electrical pulse detectors give information about nuclear radiation quickly and easily; whereas the track detectors are slow and tedious. Therefore, electrical pulse detectors are generally used for the measurement of activity from radioisotopes.

The counters that detect the nuclear radiation by the electrical pulse generated by radiation are:

1. Gas Filled Counters
   - (a) Ionisation counter (or chamber),
   - (b) Proportional counter,
   - (c) Geiger-Müller (G. M.) Counter.

2. Scintillation counter.

3. Semiconductor counter.

4. Liquid Scintillation counter.

5. Cherenkov counter.

A counting system consists of one or more of the following electronic units besides the detector:

1. *Power Unit.* The power to the counting system is supplied generally from mains. In some cases batteries can also be used. An important requirement of most of the detector is a source of stabilised (constant) D.C. supply. This purpose is served by the unit called stabilised power supply.

2. *Amplifier.* The signal (sudden change in potential is called signal or pulse) is often too small to be registered unless it is amplified. Linearity is an important requirement of the amplifier to be used with proportional counter, scintillation counter, semiconductor counter, and cherenkov counter. That is, the amplification factor of the amplifier must be independent of signal size.

3. *Timing Unit.* This can be a stop watch or an automatic unit which stops the detector at the end of a preset time, or registers the time necessary to accumulate a certain number of counts.

4. *Discriminator.* This unit rejects the pulses below and/or above a certain height.

5. *Coincidence Unit.* This unit rejects all pulses that do not reach it almost simultaneously (in coincidence). A coincidence unit is generally used in liquid scintillation counting to eliminate the noise pulses of the photomultiplier tube (see Fig. 5.7).

6. *Anticoincidence Unit.* This unit rejects all the pulses that reach it almost in coincidence. This is used for electronic shielding against cosmic radiation, pulse height analysis, etc.

7. *Single Channel Analyser.* This unit accepts pulses lying between a certain present pulse-height interval. This interval is varied to count the number of pulses of different heights. It consists essen-

tially of two variable discriminators (a lower and an upper) together with an anti-coincidence unit.

8. *Multichannel Analyser.* This is the most important unit of an advanced counting system. Roughly speaking it does the same job as a single channel analyser. But the time taken by it for determining the number of pulses of different heights is very much small compared to the time taken by a single channel analyser.

9. *Registering Unit.* This may be a scaler, a count-rate meter (visible or audible), a pen-recorder, a type-writer, etc. The result of the counting is obtained from this unit.

5.2 GAS-FILLED COUNTERS (DETECTORS)

The first three counters (Ionisation, Proportional and Geiger) are very similar in their construction and operation. They are called gas-filled counters. They are one of the oldest and simplest detectors of nuclear radiations. A gas-filled counter is a gas-filled container with a central electrode, which may be a thin wire, kept at a positive potential relative to the conducting wall of the container. Fig. 5.1 is a sketch of such a container with the electrical circuit to which it is connected. Ionising radiation passing through the counter will produce positive ions and electrons in the gas which may be, for instance, argon. The electrons will be pulled towards the positively charged central electrode, and the positive ions towards the outside wall which is at a negative potential (usually at ground potential in actual practice). The total amount

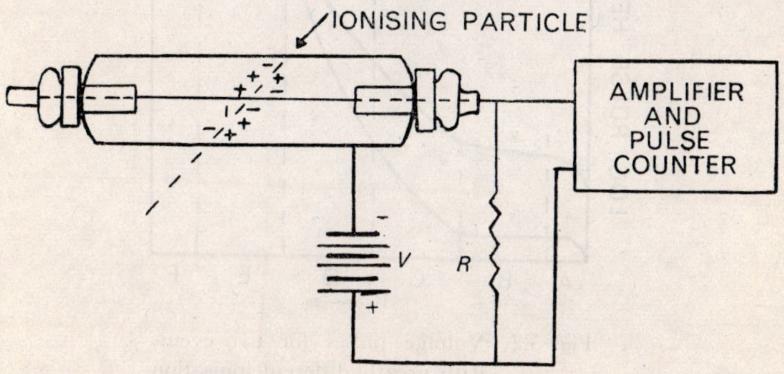

Fig. 5.1 Gas-filled counter with associated electronic circuit.

of charge that passes through the counter and the resistor R depends on the DC voltage V between the central electrode and the walls, as described in detail below.

Let us assume that the electronic circuit connected to the resistor R is designed to measure the magnitude of the voltage pulse v developed across R. Fig. 5.2 shows schematically log v *vs.* V (applied D.C. Voltage) for two different types of events in the counter. One event is the passage of a low-ionising electron through the gas and the other is the passage of a highly ionising alpha particle. In region A, the D.C. voltage V is so low that recombination of positive and negative ions takes place, so that not all the ions reach the electrodes. In region B, the voltage V is sufficiently high so that only negligible amount of recombination takes place. The gas-filled counter operating in this region is called *ionisation counter.* This region is called ion-chamber region. In region C, the voltage is sufficiently high so that electrons approaching the central wire, where the electric field is high, attain sufficient energy and produce new ion pairs in collisions with gas molecules. A limited cascade takes place, and the total amount of charge passing through R may be up to a thousand times the charge produced by the primary radiation. The gas-filled counter operating in this region C is called *proportional counter.* This region is called proportionality region. In this region the voltage pulse generated in the counter

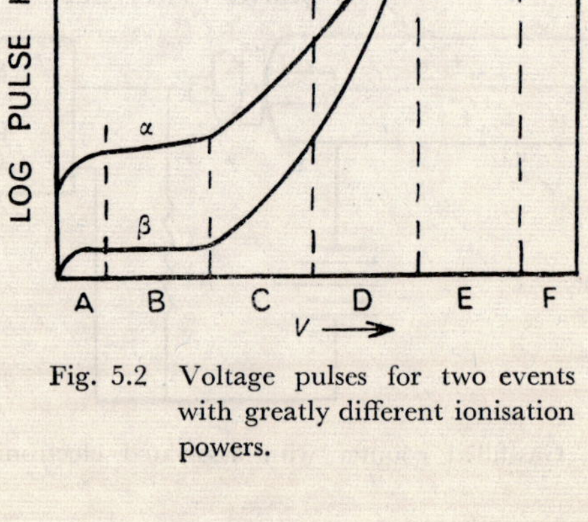

Fig. 5.2 Voltage pulses for two events with greatly different ionisation powers.

is proportional to the energy of the particle generating the pulse. For still higher values of V, in region D, the proportionality feature is lost; and in region E even a minimum ionising particle will produce a very large pulse and the pulse height is independent of the energy and type of the ionising particle. The gas-filled counter operating in this region is called *Geiger counter* or *Geiger-Müller counter* or G-M counter. This region is called Geiger region. The pulse height in this region for a typical counter is about 10 volts which can be recorded without amplification.

In G-M counter, the ionisation process spreads along the total length of the central wire, partly involving photons as intermediaries. The initial ionisation actually triggers a complete gas breakdown, which must be stopped to prepare the counter for a new event. This "quenching", as it is called, may be accomplished by electronically lowering the voltage V after each count. This lowers the speed of the counter. It is, however, usually done more elegantly in the gas itself, as described below. For the discharge to maintain itself, either one or more of the huge number of positive ions formed in an avalanche or a photon must release a new electron in the gas or at the cathode to start the process over again. When a positive argon ion, for instance, is neutralised on a metallic surface, a considerable amount of energy is released that may be used to expel an electron from the surface. Certain other gases (polyatomic) do not behave this way. If the gas is argon of 15.7 volts ionisation potential with about 10% ethyl alcohol of 11.3 volts ionisation potential, then because of charge exchanges on the way, the positive ions reaching the cathode are alcohol ions, because they have lower ionisation potential. They do not, in general, release electrons from the counter surface. The energy of the ion is used to break up the gas molecule. The alcohol is also an efficient absorber of photons and the absorption usually takes place without ionisation. Beyond Geiger region a continuous unquenchable discharge results.

The growth of the potential voltage pulse at the anode is shown in Fig. 5.3. This growth is determined by the motion of the positive ion sheath as it is repelled away from the high field region. If a second ionising event occurs before the first positive ion sheath has travelled more than about halfway across the counter, no second pulse is recorded. This is because the field close to the central electrode is still below the Geiger threshold. The time during which the counter is completely insensitive is known as the *dead*

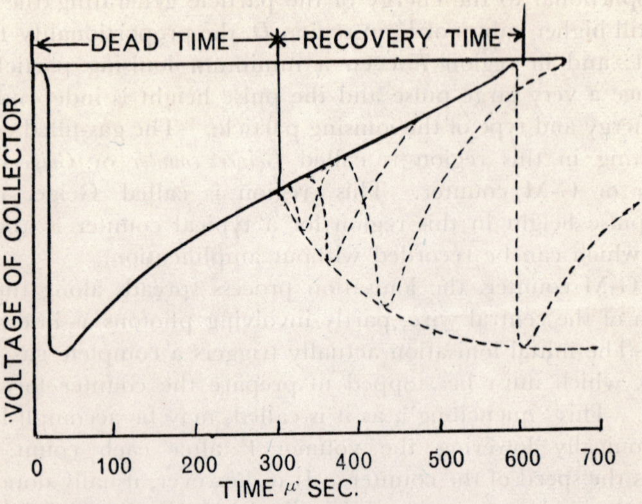

Fig. 5.3 Illustration of dead time in a typical G.M. tube.

time t_d and is typically 250 μs. This is shown in Fig. 5.3. Even after the dead time is over there is an additional *recovery time* t_r during which a smaller output pulse is produced by an ionising particle. The dead time and pulse recovery time may be as long as 600 μs and corresponds to the time it takes the positive ion sheath to reach the wall of the container. Very high counting speeds are not possible with Geiger counter.

The resolving time t of a counter is the minimum time interval between events which can be registered. For a sensitive voltage amplifier it approaches t_d. For a less sensitive amplifier t lies between t_d and t_d+t_r. If n_0 is the observed counting rate of a Geiger counter, the true counting time is reduced by n_0t units. Thus the true counting rate is given by

$$n = \frac{n_0}{1-n_0t} \tag{5.1}$$

This correction is particularly important in application of Geiger counter because the resolving times are so large ($\sim 2\times 10^{-4}$ sec).

The resolving time for a counting system can be determined experimentally by the multiple-source method. A simplified version of this method, known as the two-source method, is explained below. It involves the comparison of the counting rate indicated by the system, when two sources are counted simultaneously, with the sum of the counting rates produced by each of the sources taken individually. To accomplish this, let n_1, n_2, n_{12} and n_b be the true rate of ionising events in the counter due to source 1 plus background, source 2 plus background, source 1 and 2 plus background and background only ; let m_1, m_2, m_{12} and m_b be the corresponding observed counting rates. Then

$$n_1+n_2=n_{12}+n_b \tag{5.2}$$

or by equation (5.1),

$$\frac{m_1}{1-m_1t}+\frac{m_2}{1-m_2t}=\frac{m_{12}}{1-m_{12}t}+\frac{m_b}{1-m_bt} \tag{5.3}$$

Using the following approximate formula

$$\frac{x}{1-xt}\approx x+x^2t$$

in equation (5.3), we can get

$$t=\frac{m_1+m_2-m_{12}-m_b}{m_{12}^2-m_1^2-m_2^2} \tag{5.4}$$

Accurate measurements of t require highly accurate determination of m_1, m_2 and m_{12}, since m_1+m_2 is nearly equal to m_{12}.

An important feature of a Geiger counter is the shape and the slope of its 'plateau curve' which gives the voltage range in which the efficiency of the detector is almost constant. This is a plot of the recorded counting rate with a constant source of radiation when the voltage applied to the Geiger counter is varied. A typical plateau is shown in Fig. 5.4. The flat portion is called the Geiger region. A good Geiger counter has a plateau extending 200 volts or more and the slope of the plateau may increase by less than one percent in counting rate per hundred volts increase. This slope is partially due to the extension of the sensitive volume of the counter as the voltage increases. In addition the production of an occasional spurious pulse (pulse without particle) happens more often at high potential.

Geiger counter tubes have a limited life. This is associated with gradual decomposition of organic vapour. When positive

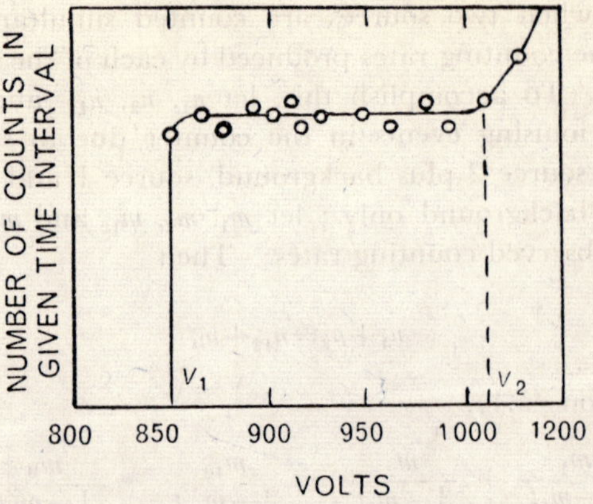

Fig. 5.4 Typical plot of number of counts versus voltage applied to a G.M. Counter.

ions become neutralised at the container wall, they dissociate into molecular fragments. Organic quenched tubes have a useful life of about 10^{10} counts. Towards the end of the useful life of a counter, gas pressure slowly increases due to the increase in the molecular fragments. This pushes up the Geiger threshold voltage and in some cases the flatness of the plateau curve is greatly reduced. This may occur with counters which use ethyl alcohol as the quenching agent; OH fragments produced by dissociation may attract electrons in the gas discharge and form heavy negative OH ions. The presence of heavy negative ions gives rise to steep plateau curves. In a halogen-quenched tube, a very small quantity of halogen gas, such as bromine or chlorine is added to a noble gas such as argon or neon.

A typical filling is about 0.1 percent of chlorine in neon. The quenching mechanism is similar to that in the organic tubes. However, there is one important difference that affects the life of the tube. In the halogen-quenched tube, the quenching gas is apparently not consumed in the quenching process. It appears that the diatomic halogen-gas molecules are dissociated in the quenching, and that there is a recombination mechanism present to replenish the supply of quenching gas. This not only extends greatly the life of the tube for normal use but also makes it possible to run the tube at higher voltages without sacrificing the life of the tube. This latter feature means that output voltage pulses of 10 volts or more can be obtained from the tube in normal operation. But the halogen vapours cannot be used with some cathode materials because of chemical action. Cathodes of stainless steel have proved satisfactory.

The halogen-quenched tubes are characterised by short plateau with rather large slopes. Typical values of plateau lengths and slopes are 150 volts and 10 percent per 100 volts, respectively. The slope in the plateau of this type of tube is primarily due to the increase in the sensitive volume of the tube with increased tube voltage. The sensitive volume of the tube is defined as that volume in which a primary ionising event will result in a discharge.

The operating voltage of G.M. tube is primarily affected by the pressure in the tube. A large range of gas pressures can be used. Most tubes have operating pressures in the range of 7 to 20 cm Hg, although pressures upto one atmosphere are used sometimes. Most of the tubes operate around 1000 volts. In many cases it is desirable that the voltage be lowered to simplify power supply. Halogen G.M. tubes have been made which operate at 250 volts.

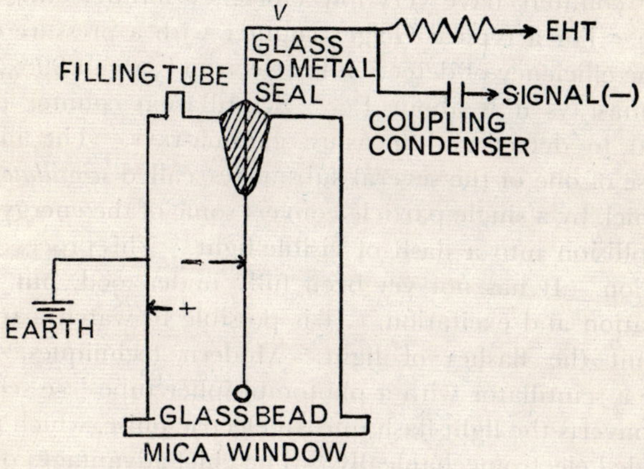

Fig. 5.5 End window G.M. counter.

To detect radiation of very small penetrating power, counters with a very thin end window are used. One such G.M. counter is shown in Fig. 5.5. Thin windows of mica, cellophane or glass of only 1 or 2 mg/cm^2 thickness will allow low energy β-particles to pass into the sensitive volume of the counters. A typical end window counter is made of copper, stainless steel or other highwork-funciton metal of cylindrical shape (6 cm length, 3 cm diameter, 0.1 cm wall thickness) with central wire (0.1 cm diameter) and filled with argon-chlorine mixture at about 10 cm pressure in the ratio of thousand to one. For detecting cosmic rays more robust counters, sometimes two or three feet long, are used.

Ionisation counter is the instrument which directly measures the gamma-ray dose. The dose is defined in terms of the number of ion pairs produced per unit volume in a gas. Proportional counters are used for a variety of purposes. For instance, as $\frac{dE}{dx}$ (rate of loss of energy) counter by which the particle can be identified, if its energy is known. In spite of the many other significant advances in the area of more sophisticated nuclear detecting methods, which are discussed below, the Geiger counter remains an inexpensive and reliable tool for the detection and measurement of nuclear radiations.

5.3 SCINTILLATION COUNTER

Geiger counters have very low efficiency for detecting gamma-radiation. For a typical Geiger counter with a pressure of 15 cm of Hg, the efficiency of detection for electron is about 99%, whereas for gamma-rays it is about 1%. Scintillation counter is usually employed to detect and measure gamma-rays. The instrument makes use of one of the several substances called *scintillators* which, when struck by a single particle convert some of the energy received in the collision into a flash of visible light. This process is called scintillation. It has not yet been fully understood, but is caused by ionisation and excitation. It is possible to watch a scintillator and count the flashes of light. Modern techniques, however, combine a scintillator with a photomultiplier tube (see section 5.4) which converts the light flash into an electric pulse, which may then be recorded electromechanically. The chief advantages of Scintillation counter are (1) it is very fast—it can give a pulse and be back in original state, ready for another, in a very short time (10^{-8} sec); (2) in many applications it can be made to give a pulse size proportional to the energy lost by the incident particle in the scintillator; (3) it has higher efficiency of detection for gamma-ray than gas-filled counters.

Fig. 5.6 is a schematic diagram of a scintillation spectrometer used in a counting system. The nuclear particle produces a flash of light in the scintillator. By means of the light pipe and reflector, a large fraction of the light is transmitted to the photocathode of a

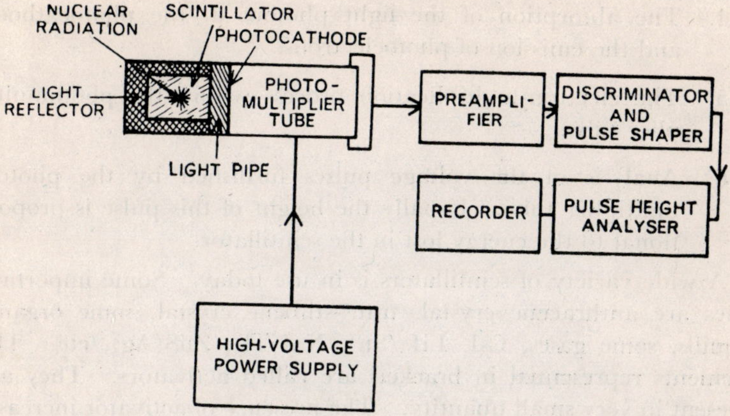

Fig. 5.6 Schematic diagram of a scintillation spectrometer.

photomultiplier tube. The photoelectrons emitted at the photocathode are multiplied many times by means of the electron multiplier section of the photomultiplier tube. The resulting current pulse produces a voltage pulse at the input of the preamplifier. This pulse, after passing through the discriminator and pulse shaper, is counted by the electronic counter. Alternatively, the electronic counter could be replaced by a differential pulse-height analyser as in the scintillation spectrometer.

For the purpose of understanding the system, one can divide the operation into six following consecutive events:

1. The absorption of nuclear radiation in the scintillator, resulting in excitation and ionisation within it.
2. The conversion of the energy dissipated in the scintillator to light energy. Among the desirable properties for good scintillator are : high efficiency for the conversion of the energy of nuclear particles to the energy of fluorescent radiation, transparency for their own fluorescent radiation, short decay time of this radiation and spectral distribution of scintillation consistent with responses of available photocathodes. Requirements for other properties such as density, form, state of matter (solid, liquid or gas), and versatility in sizes and shapes vary with the application for which the detector is intended.
3. The transit of the light photons to the photocathode of the photomultiplier tube.

4. The absorption of the light photons at the photocathode and the emission of photoelectrons.

5. The electron-multiplication process within the photomultiplier tube.

6. Analysis of the voltage pulses furnished by the photomultiplier tube. Usually the height of this pulse is proportional to the energy lost in the scintillator.

A wide variety of scintillators is in use today. Some important ones are anthracene crystal, trans-stilbene crystal, some organic liquids, some gases, CsI, LiI (Sn), NaI(Tl), ZnS(Ag), etc. The elements represented in brackets are called activators. They are present in very small quantity. The presence of activator increases the efficiency of scintillator.

Sodium iodide crystal is a very important scintillator. The light emitted in the ionisation-excitation process lies in the ultraviolet region and is not easily detected. The crystal is therefore activated with a fraction of a percent of thallium iodide which shifts the wavelength into visible region, suitable for photomultiplier detection. Large crystals can be made from sodium iodide. Almost the only drawback of this kind of crystal is that it is very hygroscopic, for which reason the scintillator must always be effectively sealed from the atmosphere. The higher density of CsI favours better absorption of the gamma-rays. On the other hand, the light output from this kind of crystal is only one quarter of the output of the sodium iodide crystal ; the CsI crystal moreover, has a longer after-glow and therefore makes a slower counter.

NaI(Tl) crystals, flat as well as well type (for higher geometry), are extensively used as scintillators in gamma-ray spectrometry. Gamma-ray spectrometry has developed into a method of analysis which offers wide scope in its practical applications. The main problem of this field is to analyse various peaks that appear in a gamma-ray spectrum. This is discussed in sections 5.5, 5.6 and 5.7.

5.4 PHOTOMULTIPLIER TUBE

Photomultiplier tube, whose schematic diagram is shown in Fig. 5.7, consists virtually of a combination of two electronic systems in a single envelope, namely,

—a photocathode, which bears a strong similarity to a photocell,

—a current-amplifying valve system operating on the principle of secondary emission.

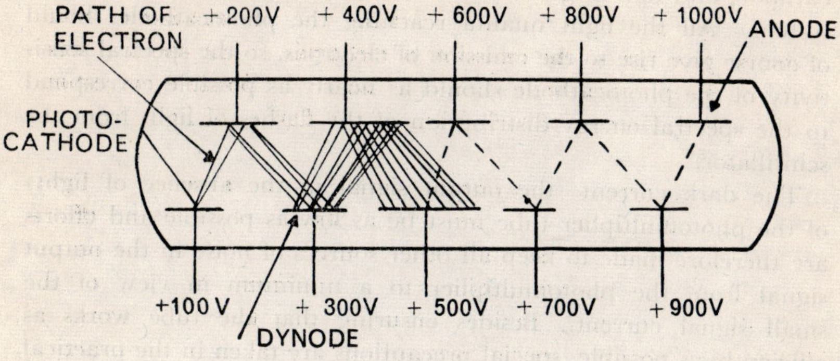

Fig. 5.7 Photomultiplier tube.

The photocathode is made of a material which exhibits photoelectric effect, that is, it emits electrons when bombarded by light quanta. These electrons will travel towards an electrode that is maintained at a positive potential with respect to the photocathode. Such an assembly of photocathode and positive electrode is almost identical with conventional photocells.

The fluctuations in the current produced by the impact of light quanta upon the photocathode are so weak, that a high degree of amplification is needed before use can be made of them and this is where the multiplier tube comes in. As pointed out, the tube operates on the principle of secondary emission. Certain materials when struck by an electron emit one or more electrons and, in certain cases, a great many.

This process is put to work a number of times in succession in the multiplier tube. A single electron ejected from the photocathode by an incident light quantum in turn liberates a number of electrons from the secondary emission electrode (the first dynode) and this, repeated ten times or more, results in an avalanche of electrons. The cascade of electrons from the last dynode is intercepted by the anode of the multiplier tube where, by reason of their vast numbers, they produce considerable fluctuations in voltage in the anode circuit. In order to ensure that none of the electrons emitted by the photocathode is lost, most multiplier tubes incorporate electron-optical or magnetic focussing to direct the electrons towards the first dynode.

The characteristics of photomultiplier tubes are dependent on the construction of the tube, the material employed for the photocathode, and to a large extent, on the efficiency of the multiplying system. All the light quanta reaching the photocathode should of course give rise to the emission of electrons, so the spectral sensitivity of the photocathode should as nearly as possible correspond to the spectral energy distribution of the flashes of light from the scintillator.

The dark current (the output signal in the absence of light) of the photomultiplier tube must be as low as possible and efforts are therefore made to keep all other sources of noise in the output signal from the photomultiplier to a minimum in view of the small signal current. Besides ensuring that the tube works as efficiently as possible, special precautions are taken in the practical application, such as cooling of the tube and, in any case, maintaining a constant temperature.

The gain obtainable from the photomultiplier tube depends to a very large extent on the voltage between the dynodes:

$$V=(C.E)^{n} \qquad (5.5)$$

Where

V = amplification

C = constant, dependent among other things on the kind of material of which the dynodes are made

E = potential difference between successive dynodes

n = number of dynodes.

A variation in the voltage E between dynodes by an amount dE will result in a change in the amplification factor of the tube equal to:

$$\frac{V}{dV} = n\,\frac{E}{dE} \qquad (5.6)$$

Stabilisation of the supply voltage to the photomultiplier tube within close limits is therefore very essential, particularly for the purposes of spectrometry. As a result of certain necessary precautions, the amplitude of the output signal from the photomultiplier tube does correspond to the energy of the gamma quanta responsible for the scintillations.

5.5 IDEAL GAMMA-RAY SPECTRUM

The ideal gamma-ray spectrum would consist of lines. The absorption of gamma-ray is caused mainly by the three following processes : photoelectric effect, Compton scattering and pair production. If monoenergetic gamma-rays of energy E_0 are recorded by a scintillator, the ideal spectrum would assume the line shape as shown in Fig. 5.8. In practice, this does not happen because of the reasons explained below.

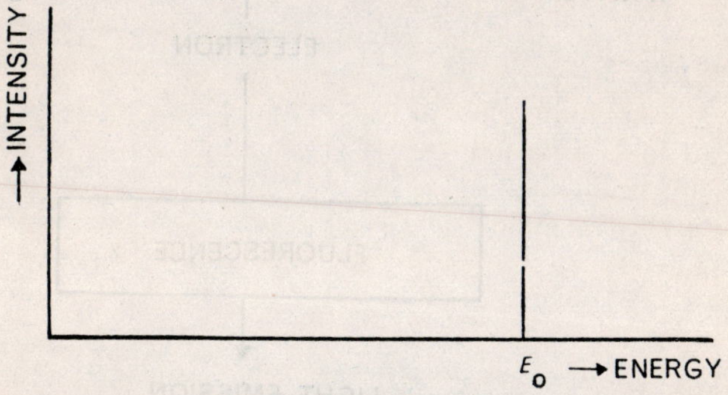

Fig. 5.8 Idealised energy spectrum of a monoenergetic gamma-ray.

What happens when a gamma-ray is absorbed by the photoelectric effect is shown in Fig. 5.9. The gamma-ray ejects an electron from one of the shells of an atom in the scintillator. The gamma-ray in this process loses its total energy, and the electron acquires energy equal to the energy of the gamma-ray minus its own binding energy. Very shortly after the ejection of the electron its binding energy is released in the form of X-rays or Auger electrons or both. If these X-rays are absorbed into the neighbouring medium of the scintillator immediately after they are emitted, a single peak called full energy peak will appear at E_0 in the spectrum. But if the X-rays escape the scintillator, a second peak will appear in the spectrum as shown in Fig. 5.10 at $E_0 - E_x$, where E_x is the energy of escaped X-rays. The probability of escape of X-rays is greater for smaller scintillators. Because of the limited resolution of the scintillation spectrometer, the escape peak is not separate from the full energy peak above 100 KeV.

The combined action of photoelectric and Compton interactions is shown in Fig. 5.11. In Compton scattering a part

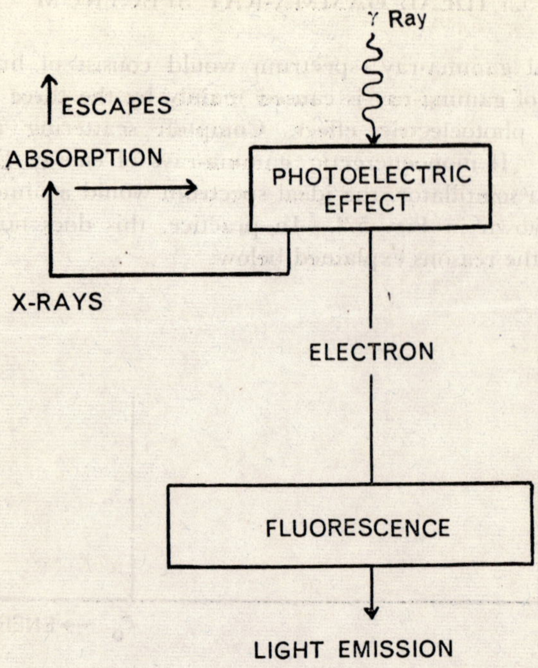

Fig. 5.9 Diagram illustrating photoelectric absorption. The X-ray quantum either escapes or is absorbed due to second photoelectric effect.

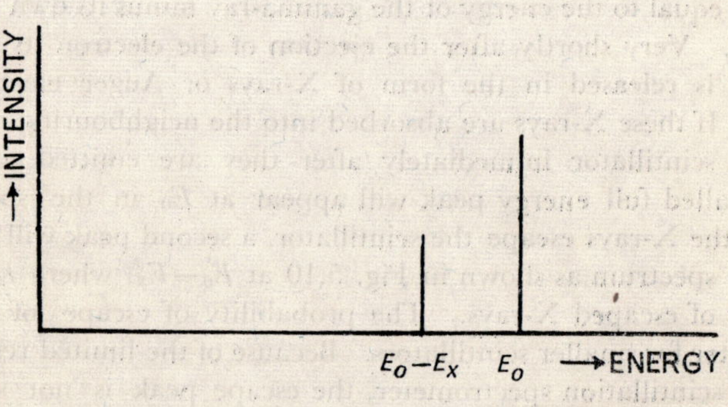

Fig. 5.10 The idealised spectrum of Fig. 5.8 showing the escape line produced when the X-ray quantum escapes from the scintillator.

of the original energy E_0 of the gamma ray is transferred to the electron. The remaining part of the energy is absorbed either by photoelectric effect or by multiple Compton scattering. If the energy of the gamma-ray is fully absorbed due to multiple Compton scattering and subsequent photoelectric effect, the full energy peak at E_0 appears in the spectrum as shown in Fig. 5.12. However, there is a marked difference between this figure and Fig. 5.10. The energy E_c transferred to the electron in Compton scattering may take any value between zero and $E_0/(1+E_\gamma/2E_0)$, where $E_\gamma=0.51$ MeV, the energy equivalent to the mass of the electron. Corresponding to the maximum energy E_c transferred to the electron, an edge appears in the gamma-ray spectrum. This is called the Compton edge.

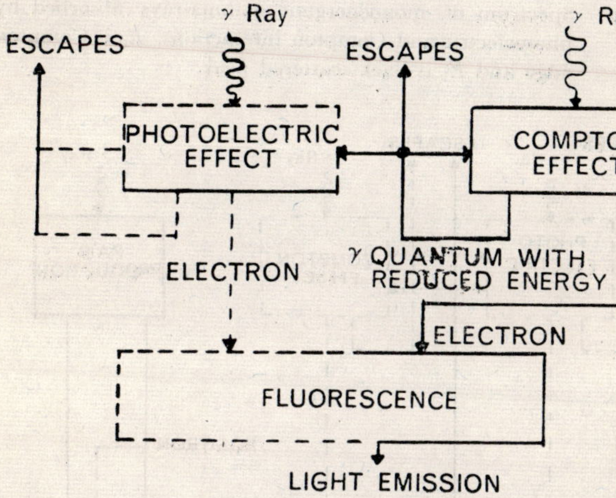

Fig. 5.11 Absorption due to combined action of photoelectric and Compton interactions.

The Compton scattering occurs not only in the scintillator but also in the surrounding material. As a result of this a back scattered peak, shown in Fig. 5.12, appears at E_t in the spectrum. E_t is the energy of gamma-ray which is returned to the crystal as a result of the Compton scattering.

In the absorption due to pair production, two particles (electron and positron) of the pair acquire energy equal to $E_p=E_0-2E_\gamma$, where E_γ is the enrgy equivalent to the mass of the electron. As shown on the right hand side of Fig. 5.13, the energy $2E_\gamma$ becomes

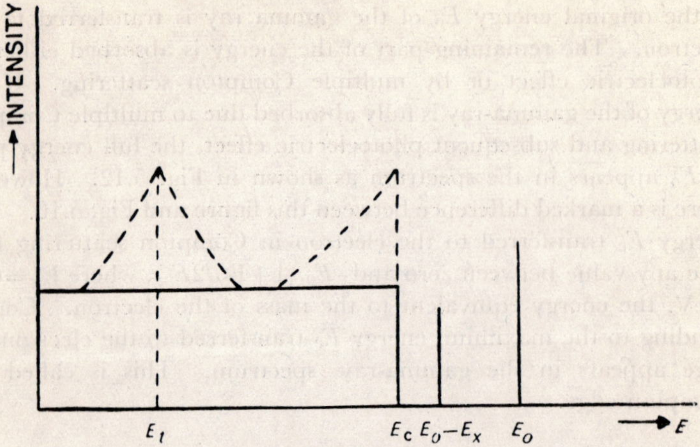

Fig. 5.12 Spectrum of monoenergetic gamma-rays absorbed by photoelectric and Compton interactions. E_c is Compton edge and E_t is back scattered peak.

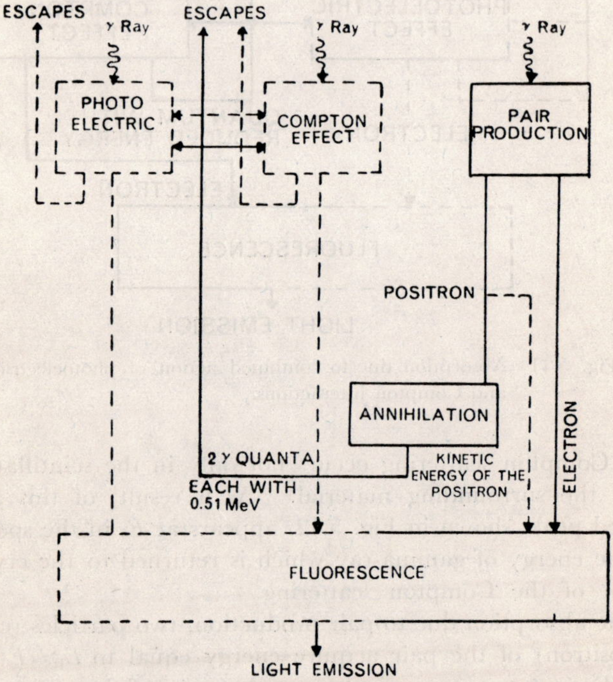

Fig. 5.13 Absorption due to combined action of photoelectric, Compton, and pair production processes.

available as two gamma quanta on annihilation of the positron. These gamma quanta interact with the scintillator material. If they are both absorbed, full energy peak at E_0 appears in the spectrum. But if one gamma quantum escapes, it gives rise to a peak called single escape peak at $E_0 - 0.51$MeV. If both the gamma quanta escape, a peak called double escape peak appears at $E_0 - 1.02$ MeV. Both these peaks are shown in Fig. 5.14.

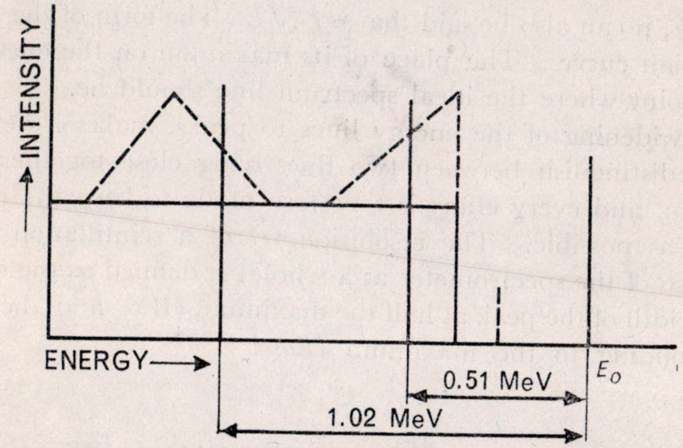

Fig. 5.14 Spectrum of monoenergetic gamma ray having undergone all three absorption processes.

5.6 CAUSES OF LINE WIDENING

Instead of the lines shown in Fig. 5.14, the spectrum will reveal 'peaks' at the corresponding points. The spectrometer records pulses to the left and to the right of the points at which the line should appear, because the amplitude of the pulses at the anode of the photomultiplier, produced by the absorption of equal energy gamma quanta by the scintillator, reveal small differences. These variations in the amplitude may be due to various causes, namely:

a. Owing to the nature of the amplification by the photomultiplier system, a spread in the pulse amplitude is found.

b. The number of electrons released in the scintillator by absorption of a gamma quantum is not a fixed quantity in relation to a given quantum of energy.

c. Besides these unavoidable causes of variation, there are also other reasons for the spread in the pulse amplitude. These are inhomogeneities in the scintillator material, in the photocathode of the multiplier tube, and in the optical system between the scintillator and the photomultiplier.

Measurement of the pulse amounts in practice to counting the number of electrons N_{anode} arriving at the anode. This involves a standard deviation (σ) equal to roughly the square root of the number counted: $\sigma = \sqrt{N_{anode}}$. As N_{anode} is proportional to the energy E, it can also be said that $\sigma \propto \sqrt{E}$. The form of the peaks is a Gaussian curve. The place of its maximum on the energy axis is the point where the ideal spectrum line should be.

This widening of the energy lines to peaks, makes it very difficult to distinguish between two lines lying close together in the spectrum, and every effort is therefore made to keep the peaks as narrow as possible. The resolution (r) of a scintillation counter (and also of the spectrometer as a whole) is defined as the quotient of the width of the peak at half the maximum ($W_{\frac{1}{2}}$) and the energy corresponding to the maximum (E_0) :

$$r = \frac{W_{\frac{1}{2}}}{E_0} \quad \text{(see Fig. 5.15)} \tag{5.7}$$

$$\% \text{ resolution} = r \times 100 \tag{5.8}$$

It is usual to indicate resolution of the instrument for the 662 keV energy peak occurring in the spectrum of ^{137}Cs (see Fig. 5.15). Clearly, the resolution of an instrument will be better if $W_{\frac{1}{2}}$ is smaller. A resolution of 8 to 8.5% for the usual commercial scintillation counter can be regarded as quite good.

In order to guarantee the effective working of the photomultiplier tube and scintillation crystal, the two are often assembled as an integral unit.

It will be obvious from what has been said above, that the intensity of a gamma-ray in respect of a given energy value is determined by counting all the gamma quanta involved in the production of a given energy peak. The area of the peak in the spectrogram is a measure of the total number of pulses, so it is the area, not the height of the peak, that indicates the intensity of a gamma-ray in respect of the particular energy.

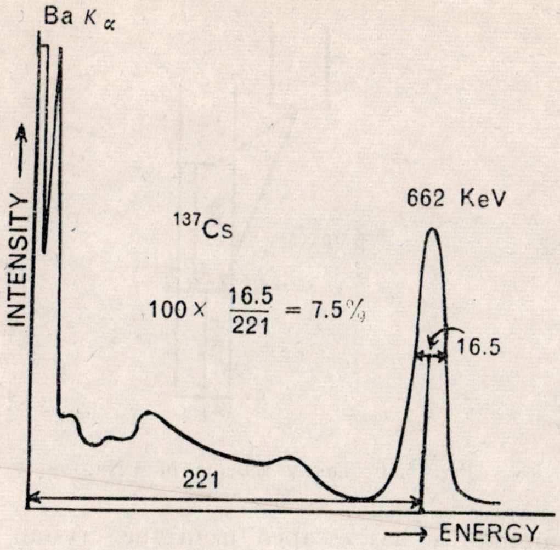

Fig. 5.15 Calculation of the relative half value resolution (r) for a given spectrometer assembly; width of the peak at half its height in cm as a percentage of the location of the centre of the peak (measured in cm).

The scintillation spectrometer has found many applications in a variety of fields. Unfortunately, as explained earlier, even mono-energetic gamma-rays produce a complex pulse-height spectrum reflecting the fact that varying amounts of gamma-ray energy may be deposited in the scintillation crystal. This is a serious limitation of scintillation spectrometer especially because of its low resolving power (about 8% for 0.662 MeV gamma-ray).

5.7 PRACTICAL GAMMA SPECTRUM

Fig. 5.14 shows an ideal spectrum; corresponding practical spectrum taken by NaI(Tl) crystal is shown in Fig. 5.16 with two gamma-rays 2.754 MeV and 1.368 MeV falling on the crystal. These gamma-rays are produced from the decay of ^{24}Na as shown in Fig. 5.17. The peak, *A*, to the far right in the pulse-height spectrum arises from events in which the full energy of the 2.754 MeV gamma-ray is deposited in the crystal, i.e., pair production with no escape or photoelectric events. The next lower peak, *B*, results from electrons produced in Compton events, where the

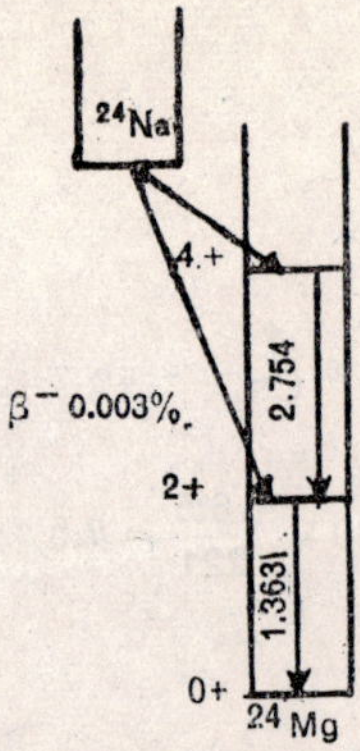

Fig. 5.16 Decay scheme of ^{24}Na.

scattered gamma-ray has escaped from the crystal. This then gives rise to a fairly broad spectrum, because the energy of the electrons depends on the scattering angle. The third peak, *C*, is caused by pair production processes in which one of the two annihilation photons has escaped and peak, *D*, arises from processes in which both of the annihilation photons have escaped. The sharp peak, *E*, in the middle of the spectrum is ascribed to photoelectric events, or pair-production events caused by the 1.368 MeV gamma-ray. The corresponding Compton peak, *F*, is also seen but the 1-quantum and 2-quanta escape peaks are not distinguishable. This is presumably because at 1.368 MeV, pair production is relatively unimportant, so that the photoelectric effect is responsible for most of the amplitude of the 1.368 MeV peak. Finally, peak, *G*, arises from 0.51 MeV annihilation photons produced outside the crystal but absorbed in it and peak *H* results from back-scattered photons from Compton events. With a strong source and nearly 4π counting a sum peak at 4.122 MeV(2.754+1.368) will also appear in the spectrum.

5.8 SEMICONDUCTOR DETECTOR

It may be noticed that in Fig. 5.16 some of the peaks (*C* and *B*) are not resolved. This is the state of affair when there are only two gamma-rays. In actual practice there may be many gamma-rays and therefore analysis of the spectrum becomes very difficult. This situation can be improved by employing detector of high

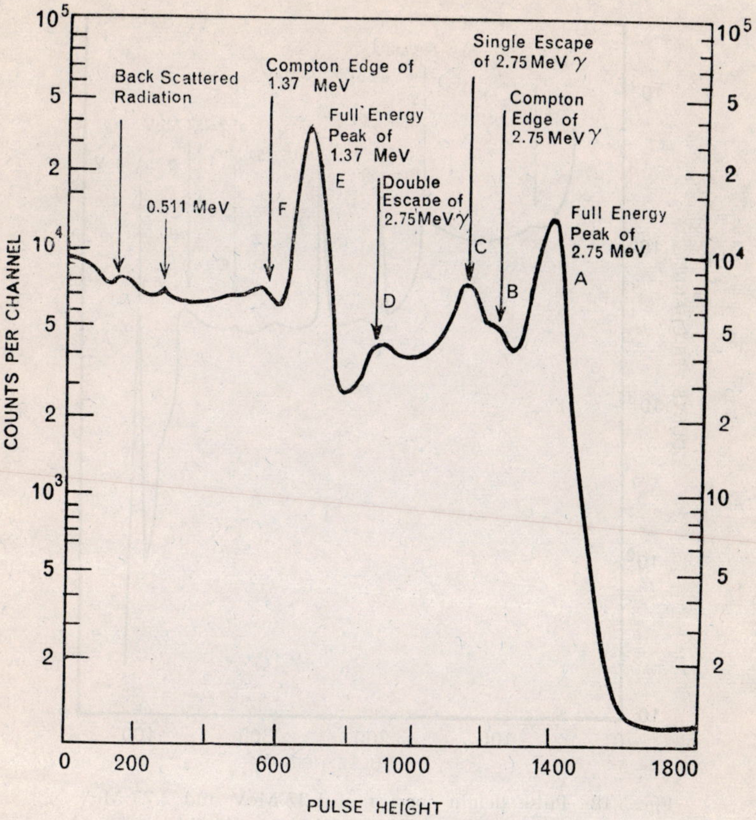

Fig. 5.17 Pulse height spectra of 1.37 MeV and 2.75 MeV gamma rays (from ^{24}Na) in a sodium iodide detector.

resolution. It is to meet the demand of high resolution that semiconductor detectors have been developed. A semiconductor detector is made up of a solid crystalline material with electrical conductivity lying between that of a good conductor, e.g., a metal, and an insulator, e.g., glass or plastic. When exposed to radiations, the electrical conductivity of a semiconductor may be increased. This property is utilised for detecting the radiation. The spectrum of 2.754 MeV and 1.368 MeV gamma-ray taken by Li-Ge semiconductor detector is shown in Fig. 5.17. This spectrum should be compared with the spectrum shown in Fig. 5.16 to see the degree of improvement in the resolution of semiconductor detector.

From these two figures it is observed that the peaks of Fig. 5.18

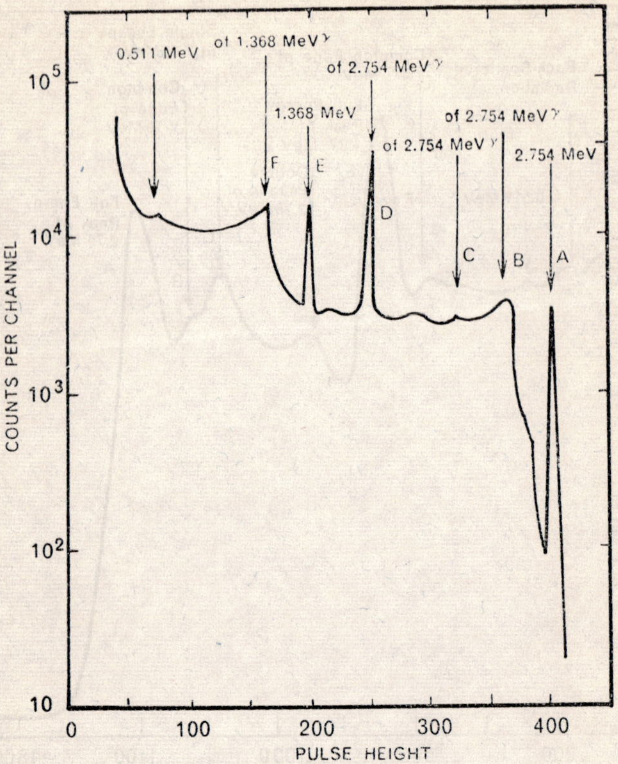

Fig. 5.18 Pulse height spectra of 1.37 MeV and 2.75 MeV gamma-rays of ^{24}Na in a lithium-drifted germanium detector. The notations have the same meaning as in Fig. 5.17.

are much sharper than the corresponding peaks of Fig. 5.17 and therefore the resolution of a semiconductor detector (Li-Ge) is much better (about 10 times) than that of NaI(Tl) scintillation counter. But the peak area which is a measure of efficiency of detection, is much larger with scintillation counter. This is because of the large size of NaI(Tl) crystal that can be grown for use as scintillation counter. Further NaI has larger stopping power than silicon and germanium which are used in semiconductor counters. The photopeak, efficiency of a lithium drifted germanium detector, having a sensitive volume of about 30 cm³, for 1.333 MeV gamma-ray, is only a few percent (4%) of 3″×3″ NaI(Tl) crystal. Still, the better resolution of Li-Ge detector is of great practical importance. Efforts are being made with success to fabricate Li-Ge detectors having efficiency similar to that of NaI(Tl) crystals.

5.9 LIQUID SCINTILLATION COUNTER

For counting low energy beta-particles from beta emitters like, ^{3}H (0.018 MeV) and ^{14}C (0.155 MeV) liquid scintillation counter is generally preferred. In this technique, the sample to be counted is mixed with the liquid scintillator (Scintillator in liquid form) so that each radioactive atom or molecule is surrounded by the scintillator molecules. This reduces the external absorption which is very serious for low energy particles. The counting sample consists of the following components:

(1) Radioactive material (^{3}H, ^{14}C, ^{35}S etc.)

(2) An organic solvent or solvent mixture (toluene, dioxane, etc.)

(3) Fluorescent material (PPO, POPOP, PBD, etc.)

The radiation emitted by radioactive material is absorbed and its energy is transferred to the fluorescent material which emits a burst of light photons or scintillation. These photons are then absorbed by the photocathode of a photomultiplier tube which converts them into electronic pulse. This pulse is recorded by the electronic system shown in Fig. 5.19. It may be noted that there are two photomultiplier tubes receiving the light photons from the sample, the output of both the tubes is passing through a coincidence unit. This is done to reduce the noise of the photomultiplier which interferes with the detection of low energy radiation. Each flash of light produced in the liquid scintillator by the emitted radiation is detected simultaneously by both phototubes. The pulses produced by each tube are amplified and fed to the coincidence circuit. The coincidence circuit rejects any pulse which does not arrive simultaneously with a pulse from the other phototube. Noise pulses are random. Thus, the noise pulses are almost rejected if the resolving time of coincidence unit is very small. The thermal noise of phototube can be reduced by reducing its operating temperature. To accomplish this a freezer is often included as a standard part of the equipment. With the availability of coincidence units with very small resolving time, the need of cooling is not important. But it is still needed for reducing thermal quenching in some cases.

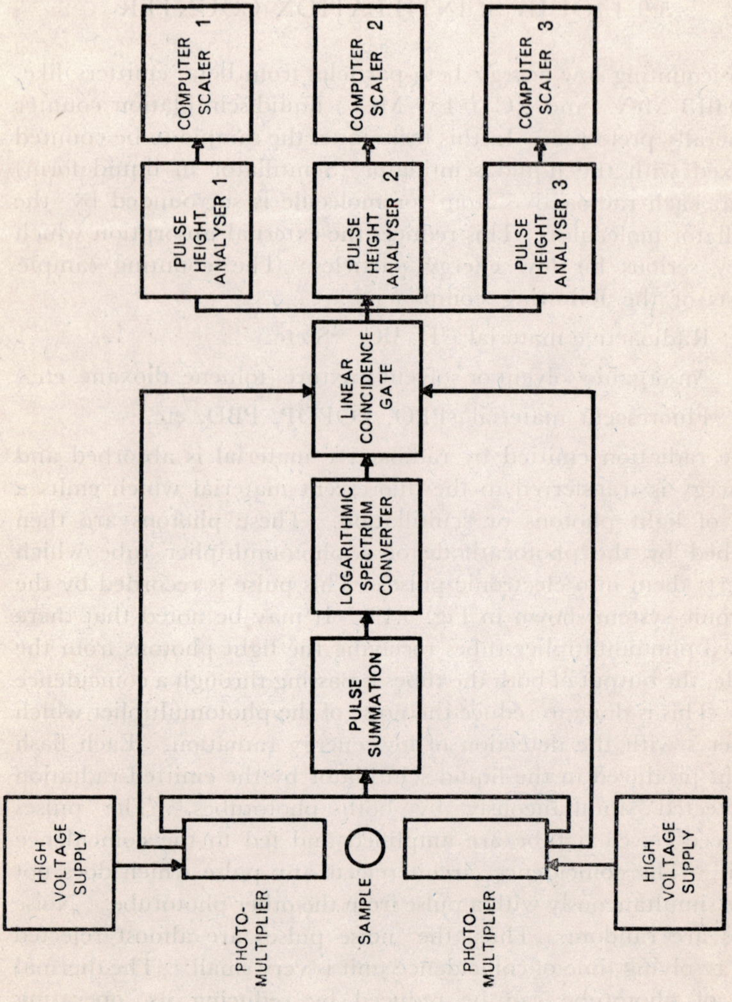

Fig. 5.19 Electronic system of liquid scintillation counter.

The container or vial used in liquid scintillation counting should contain, if possible, no radioactive material, otherwise, the background may be too high. After exposure of a container with scintillation solution to strong sunlight or fluorescent light, a high background may be observed. The decay of this background usually takes a few hours. Therefore the samples to be counted should not be exposed to such lights.

The quenching which decreases the counting efficiency is produced by processes which interfere with the production of light in the liquid scintillator and its transmission to the photomultiplier tube. Quenching is of two kinds: chemical quenching and colour quenching. In chemical quenching, compounds in solution in the liquid scintillant interfere with the transfer of energy from the emitted radiation to the fluorescent material. In colour quenching, coloured materials in the liquid absorb light emitted by the scintillator and prevent it from being detected by the photomultiplier tube. Many materials, including dissolved oxygen, act as quencher and it is nearly not possible to predict the counting efficiency in a given sample. Therefore it is necessary to determine the counting efficiency of each sample in liquid scintillation counting. Three main methods used for this purpose are; internal standard, external standard, and channel ratio.

In the internal standard method, after measurement of unknown sample activity R_s, an *aliquot* of standard solution is added to the sample and the total activity $R_t = R_s + R_{st}$ is measured. R_t is the total activity and R_{st} is the activity of the standard whose absolute decay rate A_{st} is known. The absolute activity A_s of the sample is given by

$$A_s = A_{st} \frac{R_s}{R_{st}} = \frac{A_{st} R_s}{R_t - R_s} \tag{5.9}$$

The internal standard method takes longer time. In order to reduce the time, external standard method is used. In this method a sealed gamma source is moved from a shielded storage well to a position adjacent to the sample. This method is faster but provides only relative correction. When absolute decay rates are required, the external standard source can be calibrated against an internal standard for the particular system used. The channel ratio method for quench correction is used when the activity in the sample is higher and the quenching is not too severe.

5.10 CHERENKOV COUNTER

The velocity v of light in a transparent solid or liquid is given by

$$v = c/n \tag{5.10}$$

where n is the refractive index of the medium, c is the velocity of light in vacuum. If a charged particle moving through this medium has a velocity greater than v, that is, greater than the velocity of light in the medium, light radiation is emitted from the atoms of the medium. This radiation is called Cherenkov radiation. The bluish-white light emitted from water moderated reactor core, and from transparent substances placed in the vicinity of strong radioactive source, is Cherenkov radiation. The counter that uses this radiation for detection of nuclear radiation is called Cherenkov counter.

5.11 CLOUD CHAMBER

The basic principle of the operation of cloud chamber is that a saturated vapour may not condense on a particle which is smaller than a certain minimum size, but may condense on the same size particle, if it is electrically charged. A charged particle, passing through saturated vapour, produces ionisation. Therefore many invisible particles lying near the path are electrically charged and become suitable as the centre of condensation. Condensation causes them to grow and become visible liquid droplets. Their array represents the track which is photographed for a permanent record.

C. T. R. Wilson discovered that a saturated vapour produced by adiabatic expansion of a gas, saturated with vapour of suitable liquid, will condence in a row of liquid droplets around the trail of ions produced along the track of a moving charged particle. The essential features of the Wilson expansion cloud chamber are illustrated in Fig. 5.20. A gas enclosed in a vessel of volume V_1 is saturated with the vapours of water or water-ethyl alcohol mixture at room temperature T_1. The gas-vapour mixture is expanded adiabatically by the sudden controlled movement of the piston to a volume V_2. The expansion ratio V_2/V_1 must exceed 1.31 (less for negative ions) for positive ions to act as condensation centres in an air-water vapour mixture. If this condition is satisfied, the

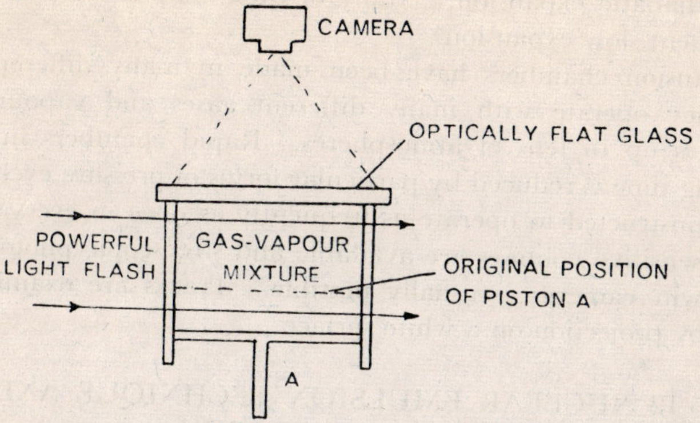

Fig. 5.20 Schematic diagram of Wilson cloud chamber.

visible droplets are formed along the track of the charged particle. The chamber has transparent windows. The visible tracks are illuminated and photographed through the wall or top of the chamber as shown in Fig. 5.20. The adiabatic expansion conveniently produced by the controlled motion of piston *A*, which forms one boundary of gas vapour volume. The expansion ratios (V_2/V_1) used in practice are limited by the appearance of background condensation on uncharged centres.

A typical sequence of operations for the photography of charged particles track is as follows. At zero time a fast adiabatic expansion occurs (in about 0.01 sec) with an expansion ratio of about 1.81. The resulting temperature drop is about 28°C for a mixture of air and water vapour and a supersaturation of 6 is produced and persists for about 0.5 sec. The interval after the production of supersaturation during which ionisation leads to track formation is known as sensitive time. During this time a burst of particles is admitted by a shutter mechanism and after about 0.2 sec. (the time for drop growth), the tracks are illuminated by a short flash of light, and photographed. The expansion ratio and timing sequence are adjusted for maximum clarity. The ions formed in the burst of particles must be removed after the photograph is taken. This is done by applying a potential of a few hundred volts (clearing field) between top and bottom of the chamber. Due to electric force ions are brought to the body of the chamber. The clearing field is removed during the expansion to minimise track broadening. Even with this field some centres for condensation may remain after adiabatic expansion. Many of these may be removed by a subsequent slow expansion.

Expansion chambers have been made in many different sizes and they operate with many different gases and vapours with pressures up to tens of atmospheres. Rapid chambers in which resetting time is reduced by particular forms of pressure cycle, have been constructed to operate as frequently as once in every 10 sec. Large sensitive volumes are available and streoscopic photography with twin cameras is usually possible. Tracks are examined in detail by projection on a white surface.

5.12 NUCLEAR EMULSION TECHNIQUE AND AUTORADIOGRAPHY

Nuclear emulsion technique is essentially a photographic technique. It records the tracks of charged particles. In optical photography, the image is formed by light whereas in nuclear emulsion the image is formed by charged particles. Photographic emulsion consists of grains of silver halides, mainly AgBr dispersed in a gelatin. The reaction of this material to light and to charged particles depends in both cases on the ionisation produced in these sensitive grains. With charged particles, ionisation results from the interaction between the atomic electrons and the field of the moving charged particle; with light, electrons are ejected from atoms by photoelectric process.

Ordinary exposures produce no immediate visible effect on the silver halide grains, but render some grains developable. Conversion of an undevelopable grain into developable grain is called latent image formation. Latent image is assumed to be a submicroscopic speck of metallic silver formed during exposure to charged particle or light.

The process of chemical development is considered to be a continuation of the process, whereby the latent image is formed. Photographic developers are weak reducing agents which require the presence of a certain critical size silver specks for their action. Electrons are transferred from molecules of the developer to the latent image; interstitial silver ions are then attracted from the body of the crystal and deposited on the latent-image speck by the electrolytic process. The developer provides a continuous supply of electrons; the supply of silver ions is replenished by the gradual solution of the surface, with the negative ions escaping into the solution and silver ions into the mass of the metallic silver. For a speck to be developable it must contain in a typical case more than about 100 silver neutral atoms. There are various methods of development and in each method there are various steps.

After development is over the emulsion is put in a fixer. The fixer dissolves all the undeveloped grains but does not effect developed grains. Due to the removal of undeveloped grain the thickness of the emulsion is reduced considerably. This is called shrinkage. After fixing, the emulsion plate is dried. The array of developed grains forms the track of the charged particle. These tracks are observed by scanning the emulsion plate by a microscope. Each field is scanned from one surface to the other in search of the event of interest.

The latent image, if not developed, is subject to fading and decay. During the time that elapses between exposure and development, the accumulation of silver in the emulsion grain diminish in size, with the result that some of the latent images are reduced to less than critical size for development centres. This process goes on in the background grains as well as in the grains traversed by charged particles, but eventually the grain spacing along a particle track becomes so large that the track is no longer distinguishable from the background, and the record of the particle is lost. The mean life of a track is approximately the time it takes for 80% of its grains to lose their latent image. Various types of emulsions are commercially available. Fading of the images, particularly in type G_5 and K_5 emulsion, does not become serious if development is completed within a month after exposure, as long as the exposed plates are stored with reasonable care. In fact, care must be exercised in storing the emulsion, both before and after exposure. Following care should be taken:

(1) Shielding from background radiation
(2) Controlled humidity
(3) Low temperature
(4) Extremes of pressure should be avoided
(5) Contact with metal should be avoided.

After processing (developing, fixing and drying), the principal hazard is excessive drying of the emulsion layer, with consequent peeling and cracking. Processed emulsion when not in use, should be stored in containers in which a relative humidity of about 60% is maintained.

Nuclear emulsion has been established as a versatile instrument, capable not only of counting particles, but giving precise information concerning their mass, mode of interaction and decay. The emulsion is a continuously sensitive detector, able to record and store information gathered over extended periods; moreover, rare events can be distinguished even in the presence of high backgrounds. Its density is much more than that of the gas in cloud chamber. Therefore, energetic charged particles can be brought to rest in comparatively much smaller emulsion. The complete histories of the particles, from their production to their end, through interaction, loss of energy by ionisation or decay, can be seen by means of the tracks recorded in the emulsion. Particles which might escape from the sensitive volume of a less dense medium are quickly brought to rest in emulsion. Because of the samll size (0.6 μ) of the developed grains of the photographic material, emulsion offers excellent angular and special resolution.

Due to the above properties and extreme simplicity emulsions have found many important applications. Nuclear emulsion is used in autoradiography. In this technique the elements under study are tagged with radioisotopes. The position of these radioisotopes in the system under investigation are indicated by the darkening of photographic emulsion placed in contact with the system. This technique has been widely used in agriculture and biology.

In general, for the faithful location of the radioisotopes in the system, the following factors should be reduced to practical minimum values:

1. Emulsion thickness
2. Specimen thickness
3. Gap between specimen and emulsion
4. Range of the radiation.

A thin specimen, such as the leaf of a plant is autoradiographed by placing a thin layer of plastic film between the emulsion and the specimen and then wrapping this 'sandwich' with black paper. The 'sandwich' is clamped together or weighted down to maintain contact during the exposure period. The plastic film retards the transfer of fluids or vapours to the emulsion which may produce images by chemical reactions between juices or vapours emitted by the specimen. Very often freezing the specimen is helpful. In this case the clamped specimen and emulsion are placed in a deep freeze, wrapped in black paper during the period of exposure. Fig. 5.21 shows the effect of geometry factors in autoradiography. For micro-autoradiography, the specimen should be labelled with low energy radiations, e.g., ^3H.

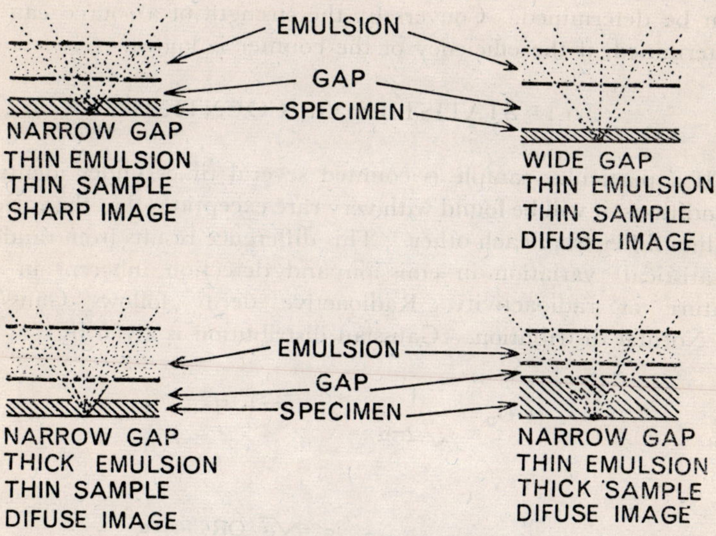

Fig. 5.21 Geometry factors in autoradiography.

5.13 EFFICIENCY OF COUNTING

Usually, not all the particles reaching the detector are counted. Some are missed because they do not produce enough excitation and ionisation to be counted. Which of them will be missed cannot be predicted, because excitation and ionisation of the counter media by the incident particle is a random phenomenon. The probability that an incident particle will be counted by a given detector is called the efficiency of the detector. It is given by,

$$\varepsilon = \frac{C}{S} \tag{5.11}$$

where ε is efficiency of the detector

C is the number of counts in a given time

S is the number of particles reaching the detector in the same time.

The number of particles *S* reaching the detector in a given time can be calculated if the absolute activity (number of disintegration per unit time) and the solid angle substanded by the detector at the source, are known. By knowing *S* the efficiency of the detector can be determined. Conversely, the strength of a source can be determined, if the efficiency of the counter is known.

5.14 STATISTICS OF COUNTING

If a particular sample is counted several times under identical conditions, it will be found with very rare exceptions that the various values differ from each other. This difference results from random (statistical) variation in emission and detection inherent in the nature of radioactivity. Radioactive decay follows Gaussian or Normal distribution. Gaussian distribution is given by

$$G_n = \frac{1}{\sqrt{2\pi\mu}} e^{-(n-\mu)^2/2\mu} \tag{5.12}$$

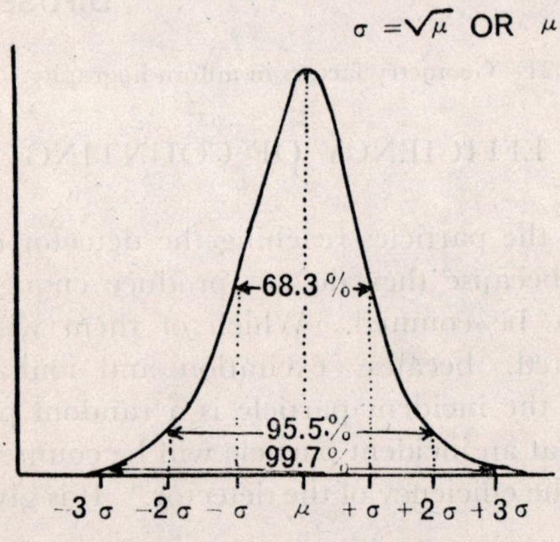

Fig. 5.22 Gaussian Distribution curve.

where G_n is the probability of occurrence of a counting rate equal to n, and μ is the mean counting rate. The standard deviation (σ) which defines the spread in counting rate is equal to the square root of the mean counting rate:

$$\sigma = \sqrt{\mu} \quad \text{or } \mu = \sigma^2 \tag{5.13}$$

Fig. 5.22 shows the Gaussian distribution curve. As indicated in this figure, 68.3% observations will be within the range $\mu \pm \sigma$, while 95.5% observations will be in the range of $\mu \pm 2\sigma$.

It is not possible to determine the exact value of μ. One can only approach μ as the number of observations are increased. In actual practice, the standard deviation is taken to be equal to the square root of the number of registered counts n, i.e.,

$$\sigma = \sqrt{n} \tag{5.14}$$

As the number of registered counts increases, the standard deviation increases but the percentage standard deviation decreases as given in Table 5.1.

Table 5.1

Registered counts	Standard deviation $\sigma = \sqrt{n}$	Percentage standard deviation $\%\sigma = \frac{\sqrt{n}}{n} 100$
100	10	10.0
1000	31.6	3.2
10000	100	1.0
100000	316	0.3
1000000	1000	0.1

It may be noted that for one percent standard deviation, 10,000 counts should be accumulated. Samples of varying strength can be counted with the same accuracy with regard to standard deviation, provided sufficient time is allowed for the same count to be registered for each sample.

If a sample gives n counts in T min., then the counting rate R of the sample is

$$R = \frac{n}{T} \tag{5.15}$$

and the standard deviation of the counting rate σ_r is

$$\sigma_r = \frac{\sqrt{n}}{T} = \frac{\sqrt{RT}}{T} = \sqrt{\frac{R}{T}} \tag{5.16}$$

Thus, the standard deviation on counting rate is inversely proportional to the square root of the counting time.

5.15 PROPAGATION OF ERRORS

When a physical magnitude is to be obtained by mathematical operations such as addition, subtraction, multiplication and division of independent observations on two or more physical quantities which include an error, the final error is given by the following equations, which are illustrated for the quantities A and B which include the errors a and b, respectively,

$$\text{Addition}: \ (A \pm a) + (B \pm b) = (A + B) \pm \sqrt{a^2 + b^2} \tag{5.17}$$

$$\text{Subtraction}: \ (A \pm a) - (B \pm b) = (A - B) \pm \sqrt{a^2 + b^2} \tag{5.18}$$

$$\text{Multiplication}: \ (A \pm a)\ (B \pm b) = (AB)[1 \pm \sqrt{(a^2/A^2 + b^2/B^2)}] \tag{5.19}$$

$$\text{Division}: \ (A \pm a)/(B \pm b) = (A/B)[1 \pm \sqrt{(a^2/A^2 + b^2/B^2)}] \tag{5.20}$$

5.16 EFFICIENT DISTRIBUTION OF COUNTING TIME

The most frequent application of equation (5.17) is in the determination of the counting rate of an active source in the presence of background radiation. The observed quantities are the background count x_2 in time interval t_2 and the total counts x_1 (source+background) in a time interval t_1. The intervals t_1 and t_2 are assumed to be measured without any error. The background counting rate C_2 and its standard deviation σ_2 are given by,

$$C_2 \pm \sigma_2 = x_2/t_2 \pm \sqrt{x_2}/t_2 \tag{5.21}$$

similarly the total counting rate C_1 and its standard deviation σ_1 is given by

$$C_1 \pm \sigma_1 = x_1/t_1 \pm \sqrt{x_1}/t_1 \tag{5.22}$$

The counting rate C associated with source is

$$C = C_1 - C_2 \tag{5.23}$$

The standard deviation σ_C of C is obtained using equation (5.17). Thus

$$\sigma_C = \sqrt{\sigma_1^2 + \sigma_2^2} \tag{5.24}$$

Hence

$$C = (C_1 - C_2) \pm \sqrt{x_1/t_1^2 + x_2/t_2^2} \tag{5.25}$$

or

$$C = (C_1 - C_2) \pm \sqrt{C_1/t_1 + C_2/t_2} \tag{5.26}$$

Suppose that a fixed time t_0 is available for counting, what is the best way to divide t_0 between background counting and total counting (source+background) so that the standard deviation of C (source counting rate) is minimum ? For this, the counting time should be divided in the proportion given by:

$$t_1/t_2 = \sqrt{C_1/C_2} \tag{5.27}$$

Using the above equation, it can be seen that if $C_1 = 4000$ Cpm and $C_2 = 40$ Cpm, then $t_1/t_2 = 10$, which means that for the most efficient distribution of counting time with minimum statistical error, the sample should be counted for a period 10 times longer than the background.

5.17 DETECTION AND MEASUREMENT OF STABLE ISOTOPES—MASS SPECTROMETER

The use of isotopes as tracers is a very powerful tool for solution of numerous problems not only in agriculture and biology but also in other branches of science. Several problems of biology could not have been solved without the use of isotopes.

Both, radioisotope and stable isotopes are used in tracer studies. The main point in favour of the use of radioisotopes is the ease of their detection and quantitative measurement by G. M. counter and scintillation counter. These detectors are easy to operate and are not very costly. The activity of as low mass as, 10^{10} micrograms of ^{32}P and 10^{-6} micrograms of ^{14}C can be measured by these detectors. Therefore, the sensitivity of the experiment using radioisotope as tracer is very high. Further, many radioisotopes which emit more penetrating radiations (gamma-ray, positive beta or orbital-electron capture) provide the possibility of studying the distribution of elements in the organism while it is still alive. This is not possible in biological work with stable isotopes as the analyses are generally performed on post-mortem specimens. On the other hand, the decay of radioisotope means that the amount of tracer is decreasing so that observation of long duration may not be possible. To a very rough approximation, a radioisotope in a given experiment can be detected up to a maximum period of about ten half-lives and, therefore, the entire experiment should not take more than 10 times the half-life of the radioisotope used in that experiment. For example, with ^{11}C which has a half-life of 20.5 min, the entire experiment including the preparation of suitable compounds for study, will have to be carried out in a period of about 3 hours. Another disadvantage of using radioisotopes as tracer is the possible harmful effects of radiations, both on the worker and on the living organism which is being studied.

A very big advantage of the use of stable isotopes in tracer studies is their permanency. It becomes necessary to use them in those experiments, where the element being traced has no long-lived radioisotope and the experiment requires long time for completion. Biologically important elements like nitrogen and oxygen do not have long-lived radioisotopes. Their longest-lived radioisotopes ^{15}O and ^{13}N have half-lives of 2.1 min and 10.1 min respectively which are too short for most of the biological experiments, even if the source of producing these isotopes is close at hand. In these cases, stable isotopes like ^{15}N and ^{18}O are employed as tracers. The main drawback of using stable isotopes in tracer experiments is that their detection is far more difficult and costly as compared to the detection of radioisotopes. Mass spectrometer, which is

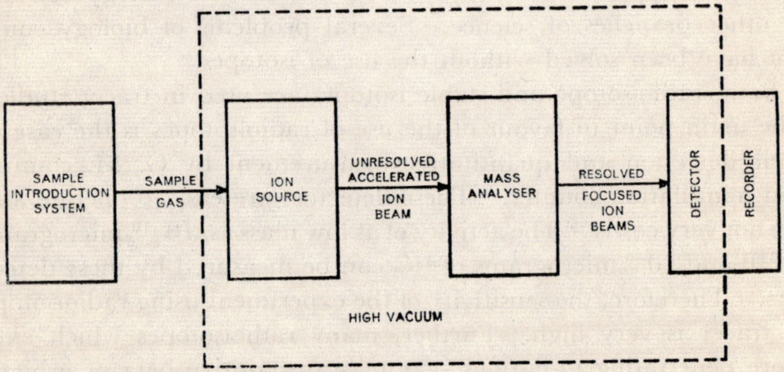

Fig. 5.23 Block diagram of mass spectrometer.

described in the following paragraphs, is usually employed for the detection and measurement of stable isotopes.

Mass spectrometer measures ionic masses (m/e ratios) and abundances of various species in a sample. Fig. 5.23 shows the main parts of a mass spectrometer : the *ion source* where a beam of ions, representative of the sample under investigation, is generated; the *mass analyser,* where various ionic masses are separated; the *detector* which detects and measures the resolved ionic masses; and the *vacuum system* which is necessary for working of all these processes are the main components of any mass spectrometer. Sample inlet system and recorder are considered as the part of vacuum system and detector respectively.

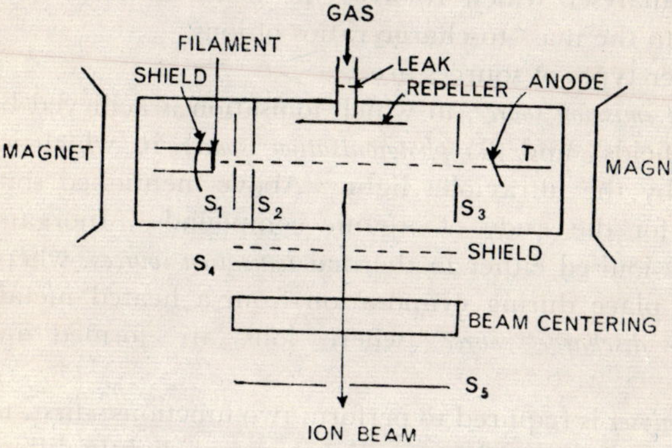

Fig. 5.24 Electron bombardment ion source.

Ion sources are of several types. The most important of them is called *electron bombardment source.* In this source the ions are produced by the collisions of neutral atoms and molecules with energetic (about 70 eV) electrons at low pressures (about 10^{-4} to 10^{-6} mm of Hg). Fig. 5.24 shows the construction of a simple electron bombardment source. The sample to be investigated is introduced in gaseous form into an evacuated chamber where it is bombarded by a stream of electrons obtained from a heated filament. Before entering the ionisation chamber through collimators S_1 and S_2 which have a narrow diameter of about 1 mm, the electrons obtained from the heated filament are accelerated by a known potential difference, applied between slit S_2 and the filament to impart them energy required to cause the ionisation of gaseous sample. The collimated beam of accelerated electrons travel through the slit S_3, towards the anode T to be collected there. A weak external magnetic field (about 100 gauss) is applied to the electron beam so as to keep the width of the electron beam small.

The gaseous sample that comes in the way of the electron beam gets ionised. The positive ions are separated from the electrons and are removed from the ionisation region by the combined action of the repeller (a plate with positive charge) and the field penetration from the accelerator section S_4 and S_5. The voltage difference between the repeller plate and S_4 directs the ions into the acceleration area, where they are accelerated to a few thousand volts by the potential difference applied between S_4 and S_5. The accelerated positively charged ion beam is then focussed on the entrance of the mass analyser which resolves the beam into its components according to the mass-to-charge ratios of ions.

The other types of sources are :

(1) *Field emission source,* in which ionisation is achieved by strong electrical fields; and (2) *photoionisation source,* in which ionisation is caused by the ultraviolet light. Above mentioned sources are employed for the study of organic compounds. Inorganic solids are usually ionised either in thermal *ionisation sources,* where ionisation takes place during evaporation from a heated metal surface or *vacuum discharge source,* where ions are formed in sparks or arcs.

Mass analyser is required to perform two functions; first, to resolve an ion beam of mass m from another beam of slightly different mass $m+dm$; and second, to maximise the resolved ion intensities. The former is called dispersive action and the latter a focussing action. There are three basic principles upon which all analysers are based; (1) magnetic analysis, (2) electrostatic analysis, and (3) time-of-flight analysis. Majority of mass spectrometers are based on the magnetic analysis principal. Therefore we will discuss the magnetic analyser in detail.

If a stream of charged particles, initially moving in a straight line, enter a magnetic field which is at right angles to the direction of motion of the charged particles, then the particles will be forced by the field to move in a circular path as shown in Fig. 5.25. If e is the magnitude of the charge carried by each particle, v is the velocity of the particle and H is the strength of the magnetic field, then the force acting on a particle which makes it move in circular path is Hev. This force is exactly balanced by the centrifugal force mv^2/r

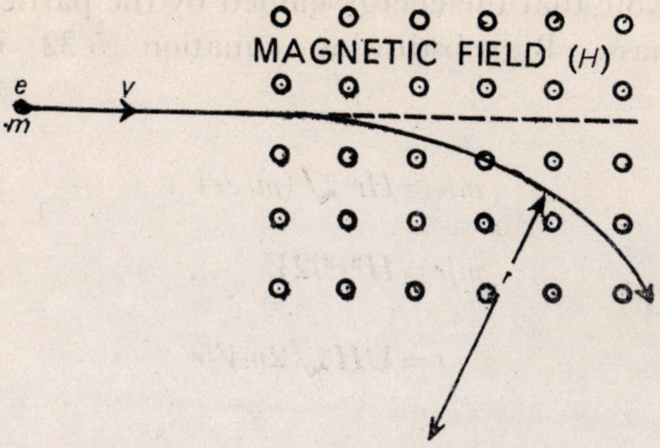

Fig. 5.25 Path of a charged particle in a magnetic field at right angles to the direction of motion.

acting on the particle of mass m moving in a circular path of radius r. Equating the two expressions,

$$Hev = mv^2/r \tag{5.28}$$

$$m/e = \frac{hr}{v} \tag{5.29}$$

or
$$r = \frac{mv}{He} \tag{5.30}$$

That is, particles having same charge and velocity but different masses will move in circular paths of different radii in a given magnetic field, which is perpendicular to the direction of motion of the charged particles. Indeed this is the basis of separating the particles of different masses in the majority of mass spectrometers.

Before entering the mass analyser, the ion beam is accelerated through a certain potential difference. The kinetic energy acquired by a particle with charge e which is accelerated through a potential difference of V volts is

$$\tfrac{1}{2}\, mv^2 = eV \tag{5.31}$$

$$v = \sqrt{2eV/m} \tag{5.32}$$

where m is the mass of the particle and v is the velocity acquired by the particle during its acceleration through a potential difference of V volts. Note that the energy gained by the particle is independent of its mass. By substituting equation (5.32) into equation (5.29) we get

$$m/e = Hr\sqrt{(m/2eV)} \tag{5.33}$$

$$m/e = H^2r^2/2V \tag{5.34}$$

$$r = 1/H\sqrt{2mV/e} \tag{5.35}$$

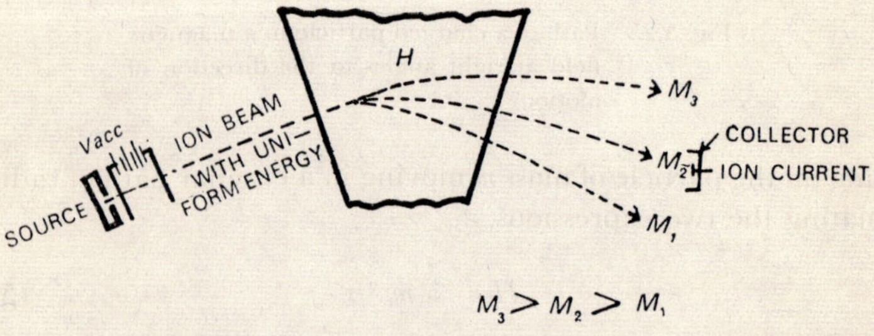

Fig. 5.26 Dispersive action of homogeneous magnetic field in a mass spectrometer. Magnetic field is normal to the plane of the figure.

This equation shows that the radius of curvature for particles of same energy but of different masses varies according to the square root of the masses. For example, a ^{14+}N ion accelerated to 2000 volts will be bent into a circular path of 24 cm radius by a magnetic field of 1000 Gauss, whereas for a ^{15+}N ion the radius will be almost 25 cm. Fig. 5.26 shows a mass spectrometer which employs a fixed accelerating voltage to obtain an ion beam of homogeneous energy and a homogeneous magnetic field to effect mass dispersion. Here the ion collector (detector) is at a fixed position corresponding to a fixed value of r. The detector is located at a place where the best focussed beam is obtained. The process by which successive ions of different masses are brought into the collector is called *scanning* the spectrum. This is done either by varying V for a fixed value of H, or by varying H for a fixed value of V. In former case, $mV/e=$ constant and in the latter case $m/eH^2=$constant.

Detection and measurement of positive ions resolved by the mass analyser can be made either by electrical or photographic means. Electrical recording follows the variation of current with time, whereas the photographic plate integrates the ion current. Ion currents in mass spectrometers are usually in the range of 10^{-8} to 10^{-16} Amperes; but in special cases, ion currents as low as 10^{-19} Amperes (1 ion/sec) or even lower can be measured.

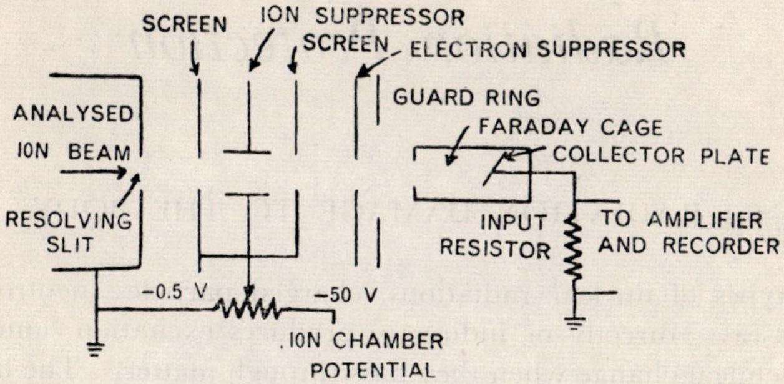

Fig. 5.27 Ion collector system.

Electrical detection is used more often than photographic method. It may be divided into three processes: (1) ion collection; (2) amplification; (3) recording. Fig. 5.27 shows an *ion collector (single)*. It consists of collector plate, and various suppressors and screens. The collector is a small metal electrode placed in an open-ended reflecting box known as the Faraday cage. The ions impinge on the metal surface which is inclined to the direction of their motion, so that the reflected ions and ejected secondary electrons cannot escape the box. In relative isotope abundance measurements, *double collectors* are often used to eliminate the serious errors caused by the momentary instrumental instabilities.

CHAPTER SIX

Radiation Protection

6.1 RADIATION DAMAGE TO THE BODY

All types of nuclear radiations, charged particles, neutrons or gamma-rays, directly or indirectly produces excitation, ionisation and chemical change when they pass through matter. The human body has not developed an instinctive reaction to nuclear radiations, as it has developed against heat. Consequently, severe radiation damage may be suffered without any realisation at the time of exposure. The cells of human body are highly specialised, each with a predetermined task to perform. Disturbance of chemical or physical balance in a cell impairs its ability to function normally. Because of the small wave length or high energy of nuclear radiation, it requires a direct hit by only a single particle to damage or destroy a cell. Therefore, it is wise to take precautions to avoid even the slightest exposure to radiation, wherever possible. Since it is clearly impossible to avoid all exposure to radiation in the present world, the International Commission on Radiation Protection has recommended safety standards based on the knowledge so far gained in the field, for the quantity of exposure that can be tolerated (see section 6.3). Before discussing these standards, it is necessary to define units in which they are expressed.

6.2 UNITS OF RADIATION

Roentgen *(r)* is defined as that dose of gamma or X-radiation which produces ion pairs carrying one Electrostatic Unit (esu) of charge of each sign per cm^3 of standard air. The charge of an electron is 4.8×10^{-10} esu. Therefore, the number of ion pairs formed per cm^3 of air by 1 r radiation is

$$1/(4.8 \times 10^{-10}) = 2.1 \times 10^{9} \quad \text{ion pairs}$$

About 33 eV is required, on an average, for the formation of one ion pair in air. Therefore, the amount of energy delivered by 1 r radiation per cm^3 of air is :

$$2.1 \times 10^9 \times 33 \text{ eV} = 6.9 \times 10^4 \text{ MeV} = 0.11 \quad \text{erg}$$

Since 1 cm^3 air at NTP has a weight of 0.0013 g, one roentgen will dissipate

$$0.11/0.0013 = 85 \text{ ergs/g}$$

Thus one roentgen is that amount of gamma or X-radiation which dissipates 85 ergs energy per gram of air. The milliroentgen (mr) which is equal to (1/1000)r is also frequently used in radiation measurements. It is often necessary to know the rate of radiation at a particular place. Therefore, radiation monitoring counters are often calibrated to read directly in mr per hour; 1 mr/hr= 0.001 r/hr.

The roentgen is basically a unit which measures the amount of X and gamma-radiations (radiation dose), regardless of the extent to which they are absorbed. That is, the amount of radiation absorbed during the exposure of a material to one roentgen of radiation is different for different materials depending primarily on the electron density of the material. Therefore, it is necessary to use a unit which is independent of the material and radiation, to express the absorbed dose. Such a unit is called RAD (Radiation Absorbed Dose). The RAD is defined as that quantity of absorbed radiation of any type (not only X-ray and gamma-radiation), which deposits 100 ergs of energy in each gram of absorbing material.

The biological effects of radiation depend not only on the amount of energy deposited (number of RADs), but also on the type of radiation. In general, particles producing a high density of ionisation (such as alpha-particles) are more destructive than the same numbers of RADs of radiation which produces only a low ionisation density. In order to compare the biological effects of different radiations, a standard of comparison, (X-rays of 200 keV) has been adopted. Based on comparison with the standard, the RBE (relative biological effect) of a radiation has been defined as:

$$\text{RBE of a radiation} = \frac{\text{Dose in RAD from 200 keV X-rays causing a specific effect}}{\text{Dose in RAD from the radiation causing the the same effect}}$$

RBE values of different radiations are given in Table 6.1

Table 6.1

Radiation	RBE
Electrons, X-rays, gamma-rays	1
Slow neutrons	3–4
Fast neutrons	10
Protons and alpha-particles	10
Nuclear fragments	20

The different RBE values are taken into account by the use of a third unit called REM (roentgen equivalent man). It is the product of the absorbed dose in RAD and the RBE of the absorbed radiation. That is, dose in REM=(Dose in RAD) (RBE). One REM of electron radiation as per definition, will have the same biological effect as one REM of gamma-radiation. Therefore, doses expressed in REM units for different kinds of radiations can be directly compared for their biological effects.

6.3 RADIATION SAFETY STANDARDS

The maximum permissible accumulated whole body dose D in REM up to an age N in years, is:

$$D = 5(N - 18) \tag{6.1}$$

According to the above formula, persons of age higher than 18 years should not exceed the average yearly dose of 5 REM and the average weekly dose should remain below 0.1 REM. The accumulated whole body dose over any consecutive 13 weeks should be less than 3 REM. When only hands are subjected to radiation, the maximum permissible levels are higher, going upto 20 REM per 13 weeks or 75 REM per year. This means that for those radiations whose RBE is 1(beta, X, and gamma-radiations), the maximum permissible average dose for the entire body (the lenses of the eyes, the blood forming organs etc.) should not exceed 0.1 RAD per week. When only hands are subjected to radiation, 1.5 RAD per week is the maximum permissible average dose. The dose in roentgen can be approximately converted into RAD, by multiplying it with a factor of 0.85. This factor has been derived from the definition of r and RAD.

According to equation (6.1), the maximum accumulated dose is zero for workers upto the age of 18 years. This is because the rapidly dividing cells are more sensitive to radiation. For a 60 year old worker, the maximum accumulated dose would be 210 REM. A dose of this magnitude received in a short period of time produces very serious effects. In fact a dose of about 400 REM received in one acute exposure is likely to be fatal.

Standards have been set for the maximum concentration of various radioactive nuclides which may be permitted in our surroundings. The relative danger of a given nuclide depends on:

1. The rate of its excretion from the body.
2. Its half-life.
3. Penetrating power of the radiation.
4. RAE of the radiation.
5. Whether the absorbed substance is concentrated into one organ or dispersed throughout the body.
6. If concentrated, the sensitivity of the organ where the radiation is concentrated.

The problems associated with prescribing the maximum permissible content of various radionuclides in our surroundings, is therefore, quite complex. Some approximate values are given in Table 6.2 for 40 hour working week.

Table 6.2

Nuclide	Tissue most affected	Maximum total body content μC	Maximum concentration in air μC per cm^3
^{14}C	Fat	300	4×10^{-6}
^{24}Na	Gastrointestinal tract	—	10^{-6}
^{32}P	Bone	6	7×10^{-8}
^{35}S	Testis	90	3×10^{-7}
^{60}Co	Gastrointestinal tract	10	3×10^{-7}
^{90}Sr	Bone	2	3×10^{-7}
^{131}I	Thyroid	0.7	9×10^{-9}
^{137}Cs	All body	30	6×10^{-8}
^{144}Ce	Bone	5	10^{-8}
^{239}Pu	Bone	0.04	2×10^{-12}
^{244}Cm	Bone	0.1	9×10^{-12}

6.4 RADIATION DOSE MEASURING INSTRUMENTS

The most important works of a radiation surveyer are: (1) measurement of radiation dose received by individual persons and (2) surveying the laboratories, equipment, clothing, hands etc. for biologically harmful radioactive contaminations. Many types of instruments are used for this purpose. The principles of their working have already been discussed in Chapter 5. Instruments that are generally used in practice for this purpose may be divided into a few categories and are discussed below.

6.4.1 Film Badges

Almost every person working with nuclear radiations routinely wears a badge-type holder of about $1\frac{1}{2}$ by 2 inches size containing photographic film, which gives a permanent record of general body radiation, integrated over a period of time, usually one week. Its principle of working is similar to nuclear emulsion technique (see section 5.12). By suitable calibration, the blackening of the film can give an indication of the number of RADs that have been absorbed by it. From this, corresponding dose received by an individual can be determined. Development, calibration, and dosage evaluation for the film badges can be obtained on a subscription basis from the Bhabha Atomic Research Centre, Bombay.

Ordinary dental X-ray film is generally used for beta and gamma dosimerty in the range 0.05 to 20 RADs. To obtain information about the type and energy of radiation, various filters (plastic, aluminium and cadium foils) are placed over certain areas of the film. Because thermal neutron flux is not well indicated on ordinary films, special boron-loaded films are used. Fast neutrons and radiations of very high energies generated by big accelerators may be monitored by nuclear emulsion. Fast neutrons produce recoil protons which leave tracks in the emulsion film. The tracks are counted under a microscope after the film is developed.

6.4.2 Pocket Ion Chambers

Film badges give the total dose received over the whole exposure period (usually one week). However, sometimes, it is necessary to know the daily exposure. Pocket ion chambers are used to get this information. Usually, pocket chamber is an air capacitor (or condenser) charged up to a known potential (usually 150 volts) before using it. The ionising radiation entering the chamber (mainly gamma-rays) causes the capacitor to discharge. The residual charge, read at the end of the day on the electrometer, gives an idea of the dose received by the wearer of the pocket chamber. Air capacitors lined with boron are used to estimate the slow neutron dose in the same manner. Pocket ion chambers are worn like fountain pens.

These pocket chambers are calibrated in roentgen units with full range of 0.1 r or 0.2 r, so as to easily detect general radiation dosage below permissible levels. They are not sensitive to soft radiations that do not penetrate the chamber wall. Some of them produce audible signal whose frequency is proportional to dose rate.

6.4.3 Portable Counters and Survey Meters

The devices discussed above give integrated dose over a period of a week, day or shorter time. For instantaneous measurement of radiation intensity (dose rate) portable G. M. counters or Survey Meters (larger ionisation chambers) are used. The ionisation chambers with compact D.C. amplifiers, operated from self-contained batteries, weighing about 1 kg (portable), are available. Models of this type have several calibrated ranges, from about 20 mr/h to 3000 mr/h. Battery operated portable G. M. counters are also available. Their full range usually varies from 0.2 mr/h to 20 mr/h. Although they are more sensitive than ionisation chambers, and give a response more nearly proportional to biological effects of the radiation. A moveable shield, to permit the distinction between hard and soft radiations is usually provided in both types of instruments. Both are useful for surverying the laboratory and its equipments for radiation contamination. Sometimes earphone is fitted with G. M. counter so that each count may be heard. This arrangement is more convenient for the rapid survey of small activity. But in its usual form, G. M. counter is not useful for very soft beta-rays, such as those from ^{14}C. Ionisation chamber is better for this purpose because very thin window can be fitted more easily in it. Portable scintillation counter is the most sensitive monitor for gamma-rays.

6.4.4 Chemical Dosimeters

Chemical dosimeters are based on chemical reactions induced by ionising radiations. There are several forms of this device, but most of them have the same essential components, namely, a chlorinated hydrocarbon, water, and an indicator dye sealed into a glass tube. Due to radiation action, hydrochloric acid is liberated from the chlorinated compound. This dissolves in the water and changes the colour of the indicator. From the extent of colour change, the radiation dose can be estimated.

Other chemical reactions that have been used in dosimetry are the oxidation of ferrous to ferric ions and the reduction of ceric to cerous ions in solution. Fricke, or ferrous sulphate dosimeter is a very useful system for the measurement of gamma dose (for details see IAEA *Technical Report Series* **119**).

6.4.5 Medical Examination

The white cell and platelet counts in the blood provide the best indication of possible radiation damage to the human body. If there is any reason to suspect that a particular person is seriously overexposed to radiation, he must soon be medically examined.

6.5 SAFE WORKING METHODS WITH NUCLEAR RADIATIONS

In experimental work, two kinds of exposures are likely to occur:

(1) External irradiation from a source outside the body.

(2) Internal radiation from radioisotopes which have entered the body.

Protection against external radiation is obtained by three different means:

(1) Distance

(2) Short exposure time

(3) Shielding

The radiation intensity from a point source falls off as the square of the distance from the source (or faster if there is appreciable absorption of radiation by air). The flux F at a distance d from a source emitting N particles per second is

$$F = N/(4\pi d^2) \text{ particles per cm}^2 \text{ per sec.} \qquad (6.2)$$

Thus, by increasing the distance, the radiation dose can be reduced considerably. For example, a source which gives a dose of 1 mr/h at 10 cm distance, will give a dose of 10000 mr/h at 1 mm distance. Therefore, even relatively weak sources must be handled with tweezers. Usually a quick check with a dose-rate meter will tell the worker at what distance the manipulation can be done safely.

Time of exposure is another important factor in working with nuclear radiations. A concentrated radioactive solution giving 1 r/h at 30 cm can be handled from a distance of 30 cm for one minute without undue harm, if the worker is doing this thing once a week for a few months. In this operation he will receive a dose of approximately 15-20 mr per week, which is below the permissible dose of 100 mr/week.

In many cases, the distance and the time factors are not enough to keep the dose within permissible limits. In such cases, shielding is absolutely essential. With alpha and soft beta emitters (^{3}H, ^{14}C, ^{35}S or ^{45}Ca) special shielding is not required because the walls of the container will absorb all particles. With strong sources of hard beta emitters like ^{32}P, a plastic sheet of 0.5-1 cm thickness may be used. A lead sheet should not be used in this case, because bremsstrahlung increases when atomic number is high. Lead bricks are used for the shielding of gamma-rays. The half-thickness (thickness of shield needed to reduce the intensity by half) of lead for gamma-rays of several energies are given in Table 6.3.

Table 6.3

Gamma ray energy in MeV	Approximate half-thickness of lead shielding in cm
0.5	$\frac{1}{2}$
1.0	1
1.5	$1\frac{1}{2}$
2-4	2

To obtain the approximate half-thickness of water, the corresponding thickness of lead may be multiplied by 10 (the density of lead is about ten times that of water). To obtain the approximate half-thickness of any material, the necessary half-thickness of water is divided by the density of that material. The attenuation factor F and the number of half-thickness n are related in the following way:

$$F = 2^n \tag{6.3}$$

or

$$n = \log F/0.3 \tag{6.4}$$

In any set up, it is necessary to check and recheck the radiation dose with a dose-rate meter before starting the experiment.

Every person dealing with radioactive isotopes should wear a film badge. Besides film badge, and especially in the absence of film-badge service, a pocket dosimeter should be used. It is desirable to wear plastic (or normal) glasses while working with hard-beta-rays, as they are found to effect the eye lenses even if the levels of radiation are far below the maximum permissible dose. The following rules taken from IAEA *Technical Report Series* **29** should be strictly observed to avoid health hazard and to obtain correct experimental results:

(1) Each person in a radioisotope laboratory should wear a *laboratory coat*, closed to the neck. This laboratory coat should be worn only in the laboratory space where the experiments with isotopes are done but preferably not in the counting rooms. Furthermore, a plastic apron should be put on when one is dealing with stock of radioactive materials. This apron should remain in the stock room.

(2) When there is a risk of hands becoming contaminated, thin surgical gloves of rubber or plastic should be worn. These gloves have to be put on in such a way that the inside never touches the outside for prevention of direct contamination of skin.

As soon as the risk for contamination of the skin is no longer present, the gloves should be removed, because they constitute a source of contamination of glassware and equipment.

A thin layer of 'barrier cream' spread on the skin of the hands is used by a number of workers in situations calling for something between rubber gloves and bare hands.

(3) To prevent contamination of gloves, hands or equipment, *paper tissues* are always used as an intermediate. After use, these tissues have to be disposed of in (foot-operated) waste bins or large drums.

(4) For avoidance of the spreading of radioactivity by means of shoes, linen or plastic shoe covers or special shoes may be used in the radioisotope laboratory.

(5) Any equipment or glassware which requires operation by mouth may *not* be used in a radioisotope laboratory. Specially designed equipment and glassware have to be used. Suction is usually applied by means of a vacuum or a plunger pipette.

(6) It is obvious that eating, drinking and smoking are strictly prohibited.

(7) For checking of gloves, hands, laboratory coats and equipment, frequent monitoring is most essential. Any contamination that is observed should be removed in the appropriate way. Be careful not to allow wet glassware etc. to drip on the detector.

(8) For prevention of contamination of benches and floors, all manipulations with strong radioisotope solutions should be carried out in *trays*. The bottoms of the trays should be covered with absorbent paper.

(9) Avoid cross-contamination by using glassware, can openers, tweezers, etc. for one particular isotope only.

(10) When work at a μC level is being done, sources of mC level should not be admitted to that laboratory. When this is unavoidable, the use of highly active sources should be restricted to a limited space only (fume cupboard).

(11) Radioactive waste should be controlled and disposed of according to the recommendations of the International Commission on Radiological Protection. Generally, liquid waste is stored in polythene containers and not disposed of through laboratory sinks. Solid waste should be put in foot-operated bins. Under no circumstances may isotopes of greatly different half-lives be mixed.

(12) Radioactive *waste* should be stored and disposed of by ordinary means only when the activity has decayed or when the waste is diluted to permissible concentration. Long lived isotopes cannot be disposed of and should be stored. In some countries a central organisation is in charge of collection, storage and/or burial of radioactive materials.

(13) Each container, beaker, vessel or bottle that contains a radioactive solution should be marked 'radioactive', and the activity concentration in the solution, as well as the nature of the isotope, and the date should be mentioned. Each container of activity should be placed inside a larger non-breakable container and should be surrounded by tissue paper; contamination will then be reduced to a minimum in case of an accident.

(14) Any operation in which radioactive dust may arise should be carried out in a *glove-box*, in which a light under-pressure can be maintained. In the exhaust system a dust filter must be present to collect radioactive particles. These precautions are particularly important in the case of α-activity.

(15) Before leaving the laboratory, check hands, clothes and shoe soles with a suitable monitor.

6.6 DECONTAMINATION

In spite of all care, contamination may occur. Soft soap may be used for the decontamination of the skin. Glassware may be effectively decontaminated by either 10% nitric acid or 2% ammonium bifluoride or chromic acid or 10% hydrochloric acid. Aluminium parts may be decontaminated with 10% nitric acid. Phosphoric acid plus a spreading agent can be used to decontaminate steel parts. Lead can be decontaminated by treating it with 4N hydrochloric acid until the reaction starts. Thereafter, it should be washed well with a dilute alkaline solution and finish with water. Wax surface of linoleum should be removed with xylol or trichlorethylene, to decontaminate it. Painted surfaces can be decontaminated by washing them with spreading agent and ammonium citrate or ammonium bifluoride. It is very difficult to decontaminate wood and concrete. Therefore, they are removed from the laboratory when contaminated.

6.7 SPECIAL FEATURES OF RADIOISOTOPE LABORATORIES

Radioisotope Laboratories are usually divided into two parts: (i) sample preparation, room, (ii) counting room. It is important that the counting room be kept free from radioactive contamination so that the low background and healthy working condition may be maintained. The counting room should have a good ventilating system. Its floor and benches should have a surface which can be easily cleaned.

The sample preparation room for handling low activities ($\sim$1 mC) may be any ordinary laboratory which has good ventilating system and fume cupboards. Its floors and benches should be covered with plastic or linoleum. However, if moderately high activities ($\sim$1C) are to be handled, the following provisions should be made in the sample preparation room:

1. Fume cupboards should have a filter for the collection of radioactive dust in the outlet.
2. All operations involving the production of radioactive dust (e.g., grinding plant samples in mill) should be carried out in fume cupboards.
3. Under no circumstances uncovered wooden or concrete

floors and bench tops should be allowed. They all must be covered with plastic or linoleum, preferably without seams, to facilitate decontamination.

4. Taps should be foot operated for prevention of contamination from gloves or hands.
5. Outlet should be provided in floor for the drainage of water.
6. Ridges and corners in which dust may accumulate and which are difficult to clean should be absent.
7. The furniture should be made of non-porous and preferably acid-resistant material to facilitate decontamination.
8. The ventilation of the room should be very good. The air should flow from least active area to high active area. The air should not be recirculated in the room.
9. Each radioisotope is confined to a particular bench to avoid cross-contamination.
10. A vacuum cleaner will prevent the spreading of dusty contamination.
11. Facilities for radioactive waste disposal as per recommendations of the International Commission on Radiological Protection (see IAEA *Technical Report Series* **29**) should be provided.

PART II

CHAPTER SEVEN

Applications of Nuclear Science in Agriculture and Biology

7.1 INTRODUCTION

Nuclear science is being used in almost every branch of agricultural and biological research. A series of publications brought out by the International Atomic Energy Agency from time to time give good account of the applications. Based on the mode of use of nuclear science, the applications can be grouped into the following four main categories:

(1) Isotopic tracer

(2) Nuclear irradiation

(3) Nuclear analytical techniques

(4) Nuclear power.

The generation of power by nuclear energy is more of an engineering application. Therefore, it is not covered in this book. The other three modes of application of nuclear science in agriculture and biology are discussed in the following sections.

7.2 ISOTOPES AS TRACER IN AGRICULTURE AND BIOLOGY

7.2.1 Definition of a Tracer or Tag

A tracer, or a tag, is a means which enables something to be located wherever it happens to be. The migratory habits of birds have been studied by affixing a small metal band on them. Such an affixing is called 'tagging' or 'labelling'. The object which is affixed is called tracer, and the object on which tracer is fixed is referred to as being 'labelled' or 'tagged'. In the above example, the tracer is a metal band. The birds on which small metal band is tied are referred to as being 'labelled' or 'tagged'. Small tags (tracers) are required to study the migratory habits of flies and mosquitoes. Radioactive isotopes are used as tag in such studies. Bacteria and even a single molecule can be identified by tagging them with radioactive isotope. A molecule can be tagged by changing its natural isotopic composition. Such a tag is called 'isotopic'. Objects like boxes, mosquitoes, paper money, etc. can be tagged simply by touching with a drop of any radioactive solution. Such a tag is called 'non-isotopic'. Mostly isotopic tracers are used in agricultural and biological research. A molecule may contain more than one atom of a particular element. If all atoms of this element are tagged, the compound is called totally labelled for that element. If the specific activity corresponding to each position is equal, the compound is called uniformly labelled.

7.2.2 Requirements of a Tracer

A tracer must be easily recognisable and must not be normally present in the system being studied. It should not interfere with the normal working of the system. But the very fact that it is normally not present may well cause this to happen. For this reason, the tracer technique was of very little use in biological research in the pre-isotope era, but the coming of isotopes revolutionised it, because the isotopes have nearly fulfilled all the requirements of an ideal tracer. For all practical purposes, the chemical behaviour of different isotopes of the same element is identical. They differ in their masses and emitted radiations. This difference in radiation and mass is utilised for their identification and quantification.

Another important requirement to be fulfilled is that it must be possible to label the compound being studied with a suitable isotope, and in a suitable position. A suitable radioactive isotope is one, whose effective half-life is long enough to enable it to last for the duration of the experiment; (effective half-life of a radioisotope in a biological system is its half-life as a result of combined effects of biological half-life and radioactive half-life; biological half-life is the time required for one half of an administered substance to be excreted from the body or from an organ or section of living tissue). Enriched stable isotopes, like ^{15}N, ^{18}O, are used as tracer if the element to be traced does not have radioactive isotope of long enough half-life. In such a case detection of the tracer is not so easy, because one has to use either a mass spectrometer or optical spectrometer for the detection and measurement of stable isotopes. In addition to half-life, the radiation properties of the radioisotope should be satisfactory. On the one hand high energy beta or gamma emitter is of little use for micro-autoradiography but, on the other hand poorly penetrating low energy radiation is difficult to detect especially from thick samples.

A suitable position for the tag is the one which will give maximum information. For example, in order to clarify an amino acid as 'glucogenic' (glycogen formers) or 'ketogenic' (lipid formers), the metabolism of the carbon skeleton must be known. Therefore, only ^{14}C labelled amino acids must be used for this purpose. On the other hand nitrogen position must be tagged by ^{15}N, if the contribution of an amino acid to the 'amine pool' is to be measured. The amino acids labelled with ^{3}H or ^{18}O must be used for the evaluation of kinetic rates of peptide bond formation. Thus, for different types of studies, different labelled materials are required. The label must be firmly attached, otherwise the only information gained from the experiment will be that the compound has lost its label. This is a problem with hydrogen isotopes (Deuterium and Tritium) because hydrogen is labile and freely exchangeable not only in the obvious positions such as $-COOH$, $-OH$ and $-NH_2$ groups but also in other groups, from which the tracer will quickly spread to a large number of compounds that have no relation whatsoever to the metabolism of the one compound that was labelled originally, thus, providing completely meaningless results.

A requirement for a particular labelling raises the question of synthesis. Tagged compounds may be made either by chemical methods or enzymatic reactions, or by making use of the versatile metabolic capabilities of micro-organism to carry out reactions, that would otherwise be extremely difficult or impossible. Biosynthetic methods of labelling is classed 'isotope farming'.

7.2.3 Advantages and Disadvantages of the Tracer Technique

The sensitivity of the isotope tracer is very high. For example, 1 mC of Carbon-14 weighs 0.218 mg and gives 3.7×10^{7} disintegrations per second, and a very small fraction of this about 10^{-4} μg which gives 20 disintegrations per second, can be easily detected. Thus one million determinations can be performed with 1 mC of carbon-14, so that although the initial purchase of the isotope may be expensive each experiment may be relatively cheap.

Another important advantage of the tracer technique is that it enables an experiment to be performed under physiological conditions in the intact plant or animal to a far greater extent than with any other technique. The technique can be used both in the laboratory and in field experiments. Isotopic tracers enable one to determine what would be impossible otherwise (section 7.2.7).

The tracer technique is a physical method. It shares the advantages that most physical methods have over chemical methods. It is often non-destructive. The sample preparation is often unnecessary or simple. Where chemical purification is required, the samples can be checked quickly to see if the activity is high enough to warrant further work; if not, they can be discarded without wasting time.

There are only a few disadvantages of the tracer technique. Specialised equipments are needed, except for purely qualitative autoradiography. A competent electronic technician is needed to maintain these equipments. The isotopic tracer technique gives no information about the parts of molecule that are unlabelled. Finally, this technique cannot be applied in some cases as there may not be a suitable isotope to give the required information.

7.2.4 Limitations of Radiotracer Experiments

Isotope Effect. Strictly speaking, it is not true that different isotopes of an element behave in exactly the same way. There might be slight difference in biological properties and metabolic fate of labelled and unlabelled compounds because there are differences in chemical behaviour of different isotopes of the same element. This is referred to as isotope effect. This arises because of the difference in the atomic weights of the isotopes. Molecules containing lighter isotope react faster than those which carry heavier ones. Therefore, differences in behaviour between isotopic and normal compounds will be observed, if one looks at reaction velocities and equilibria of reactions. This effect can generally be assumed to be within the limits of experimental error and of no significance above $Z=6$ (carbon), because the difference between reaction rates depends on the ratio of atomic weights which for ^{12}C and ^{14}C is not far from unity (14/12). But it is larger for hydrogen isotopes, particularly with tritium and hydrogen, whose masses are in the ratio 3 : 1. This is the highest mass ratio for any pair of isotopes. In spite of this difference tritium has been used as tracer of hydrogen in many very useful experiments.

In order to get a clearer picture of isotopic effect, consider the decomposition of labelled urea by the enzyme urease. 'Labelled Urea' is normal urea ($NH_2-^{12}CO-NH_2$) with a small admixture of isotopic urea ($NH_2-^{14}CO-NH_2$). Urease in the presence of water, transforms urea into carbon dioxide and ammonia:

$$NH_2-CO-NH_2+H_2O=CO_2+2NH_3$$

Since ^{12}C atoms react faster than ^{14}C atoms, the CO_2 at the beginning of the reaction will have more ^{12}C. In other words, the specific activity of CO_2 will be lower in the beginning. As the reaction proceeds, the ^{14}C atoms of the isotopic urea catch up, and at 100% reaction, CO_2 has the same specific activity again, as the original starting material. Therefore, if the reaction is not complete, a difference in the chemical reactivity between the normal and isotopic urea shows up. It will be found that more of the normal urea has been used up. However, if the reaction goes to completion, no isotope effect is observed. Isotope effect is also not observed in reversible reactions after equilibrium has been established, because the isotope effect of the forward reaction is cancelled by that of the backward reaction. A quantitative interpretation of the tracer results must take into account isotope effect, if it is not negligible. Isotope effects can be ignored unless they are intramolecular or rates of reactions are being studied.

Radiation Effect. Radiation causes ionisation and excitation, and consequently cells receiving radiation are damaged. Therefore, in experiments with radioisotopes, the question arises as to how far can we neglect the damage brought about by radiation to biological material. In the case of a plant, there must be assurance that the radiation used will not prove deleterious to plant cells or organs. Various studies have shown that there is some radiation damage to plants when it is exposed to ^{32}P activity in excess of 50 mC/g P_2O_5. One should use as low a dose of radioisotope as possible in his experiment. In fact 50 mC/g P_2O_5 activity is a very much higher level of activity than is necessary for a good uptake experiment. The radiation damage will vary with the plant species. It will also vary with the nature of radiation. Beta emitters are much less dangerous to tissue, since their radiation is readily absorbed by the aqueous medium surrounding the dissolved atom. Levels at which biological effects of radiations from tracer isotopes have been reported are:

^{131}I	0.045 μC/g of body weight in mice
^{226}Ra	0.03 μC/g of body weight
^{32}P	0.05 μC/ml of solution on mosquito larvae
^{32}P	0.8 μC/g body weight in rats
^{32}P	2.0 μC/litre of nutritional fluid for plants

Radiation damage may become serious in long term experiments with intact animals or plants exposed to high doses, or where an isotope becomes concentrated in a particular region of the body, as iodine in thyroid gland.

Radioisotopical Purity. This refers to the total radioactivity present in the form of stated radioisotope. An unknown mixture of radiocontaminants with the stated isotope would be quite intolerable in an experiment. For example, a mixture of ^{32}P and ^{33}P used in uptake study would give seriously wrong results even if the content of ^{33}P is very small, because the half-life of ^{33}P is longer than ^{32}P.

Radiochemical Purity. This states as to how much activity is in the stated chemical form taking the position of label into account. The radioisotope employed must be in a single chemical state. If this is not met, the results will be in error. For example, a mixture of phosphate and phosphite in a system will give different exchange or uptake results from one containing only phosphate because the phosphite-tagged molecules are more soluble and are not precipitated in the magnesium ammonium phosphate procedure.

Chemical Effect. The chemical containing the activity must not produce any chemical effect on the system under investigation. For example, the carrier chemical should not be toxic to the plant or animal under investigation.

Exchange Error. Radioisotopes have been used widely as tracer to study reaction rates, bond strengths, molecular rearrangements and structures. In any metabolic investigation, one must know that the phenomenon measured is not a simple exchange reaction involving no new metabolic activity. In many cases, it is necessary to eliminate exchange error in tracer studies.

Decomposition of Radiochemicals. The decomposition of radiochemicals can take place by internal radiation effect; primary external radiation effect, and secondary external radiation effect. This problem is more serious during storage. Such decomposition may be limited by several means such as storage at low temperature, avoiding frequent thawing, using the material as soon as possible after preparation. Dilution with inactive carrier reduces decomposition. But this step may be diametrically opposed to the need of the experiment in which high specific activity is needed. A solution may be diluted with an inert solid such as glass beads to reduce the decomposition and retain high specific activity.

7.2.5 Planning the Tracer Experiment

Planning a tracer experiment requires considerable time and ingenuity, because radioisotopes are expensive, and their use in larger amounts than required entails risk to human health and radiation damage to experimental system. How much activity should be used in an experiment depends upon expected dilution, method of sample preparation, likely sample size, isotope half-life, duration of counting and counting statistics. The dose used in a particular experiment should be such that the final isolated sample should give counting rate 5-10 times that of the background.

A good knowledge of the system under investigation is necessary to arrive at the correct dose for a given experiment. If the system under investigation is unknown, the correct dose is determined by repeated trials. Indeed, the isotope experiment is often performed to gain more knowledge of the system under study. The following questions might be presented to serve as a check list of potential details in the planning and organisation of a tracer experiment. Is there an isotope available with appropriate decay characteristics to solve the problems raised in the research? Is the available instrument suitable for the problem at hand? Can the experiment be performed by a method such that the data have significance within the limits required for their interpretation? Can the experiment be done with due regard to radiological safety? Can the radioactive material be easily prepared in a form suitable for detection and measurement by the instrument selected for the experiment? Are techniques available for compensating for the loss or growth of a daughter activity? Are there suitable techniques and methods to permit the separation of the materials of interest from other radioactive or inert materials? Does the tracer material represent the true state of the system under investigation?

Correct answers to these questions will lead to a good planning of any tracer experiment.

7.2.6 Isotope Dilution Technique

Isotope dilution technique is an analytical aid. It is discussed here because for getting quantitative results from many tracer experiments an understanding of the principle of isotope dilution is essential. The principle is that, if a radioactive tracer is mixed with unlabelled compound, its specific activity will be reduced. A comparison of specific activity before and after mixing enables the amount of unlabelled compound to be calculated. The technique consists of addition of a known amount of isotopically labelled compound with a known specific activity S_1 to an unknown mixture containing the same compound made up of stable isotopes. The two are thoroughly mixed to obtain a uniform distribution. After a suitable treatment the same compound is isolated in pure form. It is not essential that all the compounds be recovered from the mixture. Thus, it is possible to avoid tedious processes required for quantitative separation and carry out quick analysis. Finally the specific activity S of the isolated pure compound is measured. Now if A_1 and m be the activity and mass of the labelled compound, respectively, its specific activity is,

$$S_1 = \frac{A_1}{m}$$

When mixed with a weight M of unlabelled compound, the specific activity of the mixture will be

$$S = \frac{A_1}{m + M}$$

From this,

$$M = \frac{A_1}{S} - m$$

But,

$$S = \frac{A_2}{M_2}$$

where A_2 and M_2 are the activity and mass of the isolated pure compound

$$M = \frac{A_1 M_2}{A_2} - m \qquad (7.1)$$

Very small mass (m) is taken if the specific activity of the tracer is high. In such a case m can be neglected and

$$M = \frac{A_1 M_2}{A_2} \tag{7.2}$$

7.2.7 Some Applications of Isotopic Tracers

Isotope as a tool of biological research has been aptly compared to the microscope. A large amount of work has been done on the application of isotopes as tracers in the field of agriculture and biology. It is not even possible to mention all of them in this book. Here, we will confine ourselves to some applications in soil-plant nutrition studies, plant physiology and biochemistry, and hydrology. There is no special reason for choosing these three areas of application for the present discussion except that in the first two fields it was used early with many important results, whereas in the third field (hydrology) the application of isotopes is increasing rapidly.

Isotopes in Soil-Plant Nutrition Studies. The primary source of the larger part of man's food supply is the soil. It provides the fundamental basis for crop production and animal husbandry. The early applications of radioisotopes to studies of soil fertility and plant nutrition revealed new approaches which are resulting in improved crop production. Important contributions have come from experiments which could not have been performed without isotopes. A better knowledge gained by the use of isotopes about the movement and availability of plant nutrients and water in the soil is leading to the development of improved methods of application of fertilisers, which result in greater efficiency in their use.

The solid phase of the soil is the reservoir of most plant nutrients which must go through a series of reactions given below to produce food for man and animals.

$$M(\text{Solid}) \rightarrow M(\text{Solution}) \rightarrow M(\text{Plant root}) \rightarrow M(\text{Plant top})$$

where, M is any soil nutrient. The physical and chemical processes regulating this sequence determine the quantitative relationships between supply and utilisation. All these processes have been studied by means of isotopes. The measurement of the nutrient status of the soil and the effectiveness of the fertiliser in the soil is explained below.

If a crop has access to two sources of phosphorus with no interaction between the sources, the plant will absorb phosphorus from the two sources in direct proportion to the respective amounts available. Let A be the amount of available soil phosphorus and B the amount of available labelled phosphorus fertiliser both measured in the same unit, and the respective amounts of phosphorus per plant from the two sources be a and b, then

$$A : B = a : b \tag{7.3}$$

$$A = B(a/b) = B\ [a/(a+b)]/[b/(a+b)]$$

$$A = B[(1-Y)/Y] \tag{7.4}$$

where, Y is the fraction $b/(a+b)$ of the phosphorus in the plant derived from the labelled phosphorus fertiliser. Obviously then $(1-Y)$ is the fraction derived from the soil. If p is the total phosphorus content of the plant, then

$$p = a+b$$

and

$$Y = b/p \tag{7.5}$$

We know that specific activity is the activity per unit mass. Therefore, if we know the activity X of a sample as well as its specific activity x, the mass M producing the activity is given by

$$M = X/x \tag{7.6}$$

Thus, in the above case

$$b = D/S \tag{7.7}$$

where, D is the activity per plant and S is the specific activity of the phosphorus fertiliser. After substituting for b in equation (7.5) we find

$$Y = D/pS \tag{7.8}$$

But

$$D/p = s$$

where s is the specific activity of the plant. Therefore

$$Y = s/S = \frac{\text{specific activity of the plant}}{\text{specific activity of the fertiliser}} \tag{7.9}$$

Thus the fraction of phosphorus in the plant derived from the labelled phosphorus fertiliser can be determined by measuring the specific activity of phosphorus in the plant and the specific activity of phosphorus in the labelled fertiliser. Knowing Y, the amount A of phosphorus available from the soil can also be calculated using equation (7.4) as B is already known.

An important assumption made in the above procedure is that there is no interaction between the two sources of phosphorus. This may not be strictly valid, because the fertiliser phosphorus interacts to a greater or lesser degree with soil. With less water soluble or granulated fertilisers this interaction is not so great. Experiment should be so designed as to minimise soil fertiliser interaction in the measurement of A-value.

Isotopes are being used to answer practical questions, such as how much nutrient is available in the soil. When, how, where, in what form, and how much fertiliser should be applied for its maximum uptake by the fieldgrown crops. The isotopic tracer is the only method to measure directly the uptake from the fertiliser. It may be said that isotopes are the primary means for directly measuring quantities and rates of processes involved in the movement of nutrient ions, from soil through the chain of events, resulting in the performance of a metabolic function by the ion in the plant.

Isotopes in Plant Physiology and Biochemistry. Isotopes as tracer have been used as a systematic and not as an occasional tool in biochemistry and physiology. Using heavy oxygen ^{18}O water, it was shown that O_2 liberated in photosynthesis comes from photodissociating water, rather than from carbon dioxide as asserted by previous theories. Several other features of photosynthesis have also been understood using ^{14}C. In fact, the first few years of the application of tagged atoms to the exploration of photosynthesis brought more successes than the entire 180 year preceding period.

The tracer technique has brought drastic change in our concept of the rate of biological processes in plants. It is now known that in spite of the outward immobility, the reaction and internal distribution of substances in plants proceed at rates comparable to the rates of the processes occurring in animals. The movement of assimilants along the phloem proceeds at a rate of upto 150 cm/hour or even faster. Experiments with $^{1}H_2{}^{18}O$ and $^{3}H_2{}^{16}O$ led scientists to conclude that the mobility of water in plant tissues is so great as to hamper an accurate determination of water compartmentation in the cells.

The impact of tagged atoms is the greatest in determining the biosynthesis patterns for different substances, as well as the sequence of metabolic reactions and the alternate ways of metabolite utilisation. As a result of these studies, it is now known that plant metabolism is a dynamic self-regulating system. Thus, it may be said that labelled atoms have become an essential technique for investigating the intimate organisation of life processes. But this method is fully effective only when applied in combination with other potential techniques.

Isotopes in Hydrology. Water is the blood of agriculture. The need for water is increasing because of intensified agriculture and increasing industrialisation, which are essential for raising the standard of growing population. Great efforts, therefore, have to be made to develop available water resources. Isotopic techniques are being used in the study of hydrological problems in the early stages of water resources development, where they provide information more rapidly than other conventional techniques. At more advanced stages of development, isotopic techniques are complementing conventional geochemical and geophysical techniques, and assist in the more complete assessment of the parameters which enter into water budget considerations.

Various dyes have been employed in the past for tracing water. But they are often absorbed by ground and hence give only qualitative result. Tritium is an ideal tracer for water, as it can be incorporated in the water molecule. The fractionation effect discovered in several stable isotopes present in water, especially ^{2}H and ^{18}O, are proving very useful as tools for obtaining useful hydrological information on the origin of water. The cosmic-ray produced radioactive isotopes, especially tritium, have been used for the determination of the age of groundwater. This information is of great value for estimating the recharge rates of groundwater bodies. At present, the use of isotopes in hydrology is rapidly increasing.

7.3 NUCLEAR IRRADIATION IN AGRICULTURE AND BIOLOGY

7.3.1 Radiation Effects

Nuclear radiations (alpha-ray, beta-ray, gamma-ray, neutron, proton etc.) are ionising radiations. Their radiation effects on biological systems are seen over a wide time scale from a fraction of a second to many years. These effects are summarised in Fig. 7.1. Radiation may damage biologically important molecules by

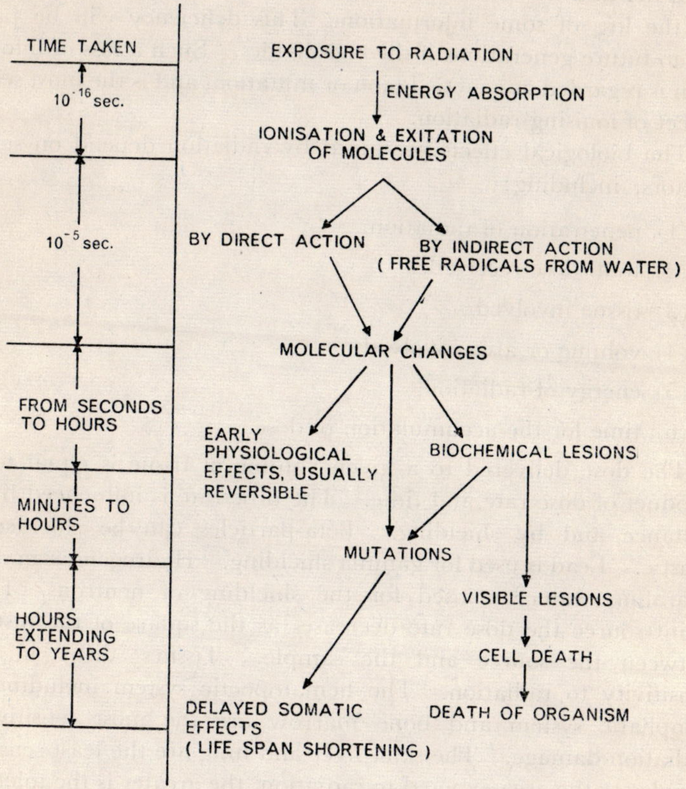

Fig. 7.1 Time scale effects of ionising radiation on living organisms.

direct or indirect action. In direct action the molecule gets altered through the deposition of energy in it as a result of direct hit with the radiation. In indirect action, the primary action of radiation is on water. The biological molecules are then attacked by highly reactive free radicals that are formed in water by radiation. A biological system suffers mostly indirect action because it is mostly water. The initial radiation lesion is essentially chemical in nature. Its manifestation as a cellular effect will depend on the types and number of molecules affected and to some extent on their location in the cell. Cellular effects of radiation can be divided into three broad groups (1) effect on vegetative function, (2) effect on information storage and (3) effect on cell division. The effect on information storage is caused by the radiation damage of DNA and hence to chromosome. The damage to DNA results in the loss of some information. This deficiency will be passed on to future generations if the cell divides. Such a loss of information is regarded as genetic lesion or mutation, and is the most serious effect of ionising radiation.

The biological effects produced by radiation depend on several factors, including:

(1) penetration of radiation,

(2) total dose received,

(3) tissue involved,

(4) volume or area involved,

(5) energy of radiation,

(6) time for the accumulation of dose.

The dose delivered to a given volume of tissue is equal to the product of dose rate and time. The dose rate is influenced by the distance and by shielding. Beta-particles can be shielded by plastic. Lead is used for gamma shielding. Hydrogenous material (paraffin, water) is used for the shielding of neutrons. For a point source the dose rate decreases as the square of the distance between the source and the sample. Tissues differ in their sensitivity to radiation. The hematopoetic system including the lymphatic system and bone marrow are the most sensitive to radiation damage. The skin, liver and lung are the least sensitive. The lesser the area exposed to radiation, the greater is the tolerance to a given dose of radiation. For example, total body exposure to 1000 RAD would almost certainly result in the death of the individual. But a similar dose restricted to a hand or foot would not result in death. If a given dose is delivered over a long period, it will cause less damage than when the same dose is given in a short period of time. Several theories like, target theory, diffusion theory and fuzzy-target theory have been proposed to explain the observed effects of radiation on living systems. Since biological systems are so complex and varied, that it has not been possible so far to develop a single unified theory, which can explain all phenomena for all systems in a satisfactory way.

7.3.2 Mutation by Irradiation

In the broadest sense, any change in the hereditary material is called mutation. Mutations are of two types: (1) chromosome mutations and (2) point mutations. When cells are irradiated, scissions in chromosomes are observed during metaphase. Most of these breaks undergo spontaneous repair by joining at the broken end. But if the radiation dose is large, the number of broken chromosomes increases and the probability that fragments of chromosomes may not combine also increases. One can observe such chromosome abberations under a microscope. Such changes are called chromosome mutations. It is caused by a gross change in chromosome, without any change in individual genes. A change in the number of chromosomes present in the cell is also referred to as chromosome mutation. Such changes are fairly common in plants, but they are rare in mammals and are fatal if they do occur. Point or gene mutations are more common. They involve a change in the genetic code in one of the four ways: (1) addition of a nucleotide to the chain, (2) removal of a nucleotide, (3) transposition of the code or (4) replacement of one nucleotide for another. Both chromosome and point mutations can occur in somatic cells. Such mutations are called somatic mutations. These may produce some effect that is apparent in the individual but is not passed on to succeeding generations in sexual reproduction. If the mutation takes place in a germ cell, it is called genetic mutation. This is passed on to later generation by sexual reproduction.

All point mutations are not immediately detectable, because a great majority of them are recessive, and must become homozygous before they can express themselves. A recessive gene may remain latent for a number of generations, until the occasion arises for the union of sperm and egg cells containing the same gene. But even those mutations which are not detected by visible phenotypic change, influence the viability of the organisms in some indirect way.

Mutations are often deleterious to the plant or animal in which they occur. This is because, during the long period of evolution both plants and animals have already tried most of the genetic possibilities for improvement. Only few possibilities for genetic improvement are now available, especially for highly developed plants and animals. That is why only a few mutations are beneficial. It is by breeding and inbreeding of these beneficial mutant strains that there have been obtained improved plants and animals which are useful to man. In the past, the procedure was to wait for a desirable mutation to appear spontaneously, as they do from time to time, and then by crossing and selection develop a plant in which the change is permanent and beneficial.

Mutations can be induced by treating the plants and animals with ionising radiations and some chemicals. The radiations and chemicals that can induce mutations are called mutagenic agents. Different mutagenic agents produce different mutation spectra. They are presumed to cause disturbances among the molecules, resulting in either the destruction of some unit or in chemical re-organisation. The mutation frequency is generally proportional to the dosage of irradiation. It has been shown that such factors as oxygen tension and temperature which are associated with irradiation, may significantly alter the frequency of mutation. Low oxygen content results in decreased mutations. The oxygen tension is less effective under intense irradiation.

Induced mutation technique has proved to be of considerable value in crop improvement programmes. Until the year 1969, at least 77 officially released varieties were produced by induced mutation in a great number of crops throughout the world. Out of these, about 60 were bred between 1965 and 1969. Induced mutation has special value in a situation where selection and hybridisation-selection procedures have been applied for a long time with the result that further progress by these methods becomes very difficult. Introduction of mutant gene may be useful in such cases. Variation in any important character can be produced by induced mutation. The main drawback of mutation is that it is random in nature. Because of this randomness and variety of wanted changes, a mutation breeding programme will have to depend for its success primarily on the efficiency of screening method used for evaluating the variable characters. The demand on the efficiency of screening technique will increase still more, if the desired change is negatively correlated with other desired characters.

Mutation in polygenically controlled quantitative traits can be utilised to improve agronomically excellent varieties in a desired respect like yield, kernel, size, earliness, quantity and quality of oil and protein etc., provided mass screening techniques are available. As discussed in section 7.4, nuclear science has already started offering such mass screening techniques for rapid and non-destructive identification of such plant characters like protein and oil, which cannot be evaluated by conventional methods on such large scales as required in successful breeding programmes. The irradiation and other mutagenic treatments of seeds are usually very useful in seed propagated annuals and biennials. Due to low metabolic rate and the absence of DNA sysnthesis in resting seeds, a variation in the delivered dose rate does not play any important role in inducing mutations. Ultraviolet light is of potential value for inducing mutations in single cells. Neutrons, gamma-rays and X-rays which are penetrating radiations, are used for homogenous irradiation of tissues. Thermal neutron irradiation of materials containing boron-10 leads very effectively to drastic mutations. Fission neutrons of around 1 MeV average energy are 10-150 times more effective than ^{60}Co gamma-rays. The use of radioisotopes, such as ^{60}Co or ^{137}Cs, in a gamma-garden for continuous irradiation during the whole or parts of vegetation period has had little value in practical breeding work. Search is on for mutagenic agents which can induce specific mutations, i.e. mutations in wanted directions.

7.3.3 Prevention of Food Spoilage by Irradiation

FAO has estimated that one-fifth of the world food crop is destroyed by insects, micro-organisms and pests. These food losses are the greatest in tropical areas, where products rich in animal protein, i.e., meat, milk, eggs and fish, decompose in a matter of hours. If this great spoilage could be reduced significantly, then millions of people could be better fed. The harmful effects of ionising radiations on living things can be used in preventing food spoilage. DNA synthesis and hence cell division is very effectively impaired by ionising radiations. This impairment of cell division is caused by a denaturation of bacterial protein. Given the right doses this impairment is obtained without serious effects on resting or much less actively dividing cells of food itself. Therefore, bacteria, micro-organisms and insect gametes may be prevented from reproducing, thus securing stability of the food.

Since irradiation by ionising radiations does not lead to an appreciable warming up of the irradiated product, it is the only treatment that may be applied through any type of packing material including those which cannot withstand heat such as paper, plastics and wood. This means that radiation can be applied after packing, thus avoiding recontamination or reinfestation. The dose distribution is more uniform in the case of irradiation than in heating. The capital cost of the food preservation by irradiation is high. But its running cost is low. Therefore, it is applicable where large quantities of food are involved. The application of radiation in the prevention of food spoilage may be considered under the following five headings:

1. Complete sterilisation, i.e., the total destruction of all bacteria, micro-organisms and spores, for long-term storage;
2. Pasteurisation or partial sterilisation, when vegetative forms of organisms are inactivated, although spores may still remain;
3. Disinfection, i.e., the inactivation of certain pathogenic micro-organisms;
4. Disinfestation by sexual sterilisation of insects;
5. Inhibition of sprouting of stored root crop.

Apart from affecting living organisms in food, irradiation may induce radioactivity in it. Heavy particles like protons, neutrons, deuterons, etc. induce long-lined radioactivity in food. Therefore, they cannot be used for food preservation. Gamma-radiation does not induce long-lived activities if its energy is less than a certain threshold. The radioactivity induced by ^{60}Co gamma irradiation of most elements commonly found in foods have short half-lives, except iodine-126 whose half-life is 13 days, which is formed in (γ, n) reaction on iodine-127. For foods containing iodine, a minimum time of storage between irradiation and consumption must be specified to allow iodine-126 to decay away. Activity induced by electron irradiation has been shown to be well below the maximum concentration permitted in food except with carbon and chlorine. The half-lives here are short, (20 and 33 min) so that they do not present any problem.

High radiation dose needed for complete sterilisation may damage the food of low specific heat by its heating effect. Biochemical effects are also produced by high radiation dose. Most amino acids are affected by high radiation doses. They are thought to be a possible source of objectionable odours and flavours after irradiation. Although it has been found that the biological value and digestibility of foods are unaffected by radiation, a lot of research is needed to clarify various biochemical effects of food irradiation.

The dose required for efficient sterilisation is about 2.5×10^6 RAD. But such a high dose of radiation makes the food unacceptable for human consumption. On the other hand the sterilisation of food wrappings and packages is quite acceptable, because they are not eaten. The shelf life under refrigeration of all types of fish can be extended upto 1 or 2 months by a dose of 5×10^5 RAD without affecting its food quality. But this is pasteurisation rather than sterilisation. With food, pasteurisation shows more promise than complete sterilisation, because it requires smaller doses. The same is true of disinfestation and to some extent of disinfection. The doses required for grain disinfestation and disinfection are about 0.5×10^5 RAD and 5×10^5 RAD, respectively. A dose of 10^4 RAD has been found to prevent potatoes from sprouting, and lengthen their storage life considerably. At present, much attention is being paid to a combined radiation and chemical or heat treatment to prevent food spoilage.

7.3.4 Pest Control by Irradiation

The application of ionising radiation for pest control is based on the fact that a comparatively small dose of radiation will make the pests sterile. They will be able to mate but the union will be unfertile and no offspring will be produced. The use of sexually sterile males (sterile male technique) is a very promising technique for selective eradication of certain specific pests in the field without much interference with the ecosystem.

In this technique a very large number (several millions) of male insects are bred in the laboratory and are made sterile by a suitable dose of irradiation. They are then released into the area where control is desired. There, they mate with the indigenous females who then produce no offspring. The success of the method depends on the fulfilment of several conditions. It must be possible to raise very large number of insects so that the sterile males will compete effectively with males of normal population. The sterilisation must have no adverse effects on the mating behaviour; and the large increase, though temporary, in the insect population resulting from the release of the sterile males must cause no serious harm to the crop or livestock. The insect should not be migratory because, while it may be easy to clear an area of the pest, it will be useless to do so, if the area is repopulated by normal insects moving in from neighbouring regions. There are several cases where these conditions have been successfully satisfied.

The eradication of the screw-worm fly in the southeastern United States by the sterile male technique led to investigations of the possibility of controlling in the same manner other insect pests, such as fruit flies, corn borers, gypsy and codling moths, the tsetse fly, the anopheles mosquito, etc. In these cases, the problem has proved to be more complex than those encountered with the screw-worm fly. From the experience gained so far with this technique, it appears that sterile male technique will be practical in the control of insects with low population. In high insect population, this technique can be used in combination with chemical and biological methods of pest control. That is, the pest population should be lowered by chemical or biological methods before the application of sterile male technique in a particular area.

7.4 NUCLEAR ANALYTICAL TECHNIQUES IN AGRICULTURE AND BIOLOGY

7.4.1 Activation Analysis

Activation analysis is a nuclear technique of elemental analysis. An ideal analytical technique should be accurate, precise, specific, non-destructive, fast, economical and universally applicable. Activation analysis satisfies nearly all the above requirements. It has gained much importance in the field of trace analysis, as it is the most sensitive method, known at present for detecting a majority of the elements. The lower limit of detection for most analytical techniques is about 10^{15} atoms. But for activation analysis it is 10^9 atoms (a million times less) for certain favourable cases. This technique is being used in a variety of fields such as agriculture biology, medicine, geology, criminology, electronics, astronautics, metallurgy, etc.

Basic Principle. Each stable isotope emits its characteristic nuclear radiation on being bombarded by nuclear projectiles. These radiations are different for different isotopes. In other words, these are characteristic radiations. In activation analysis, the elements, or more precisely the stable isotopes of a sample, are determined by means of these radiations. The sample is bombarded by nuclear projectiles of known energy and type (neutrons, charged particles and gamma-rays), produced in a reactor or accelerator. The characteristic nuclear radiations are recorded by a nuclear detector. The elemental composition of the specimen is then determined by analysing the recorded radiations. The basic principle of this technique can be understood by considering the bombardment process. The bombardment process can be represented by the following expression:

$$X+p \rightarrow \begin{cases} X \;+p & \text{elastic scattering} \\ X^*+p & \text{inelastic scattering} \\ \left.\begin{array}{l} Y \;+b \\ Z \;+c \\ \text{etc.} \end{array}\right\} & \text{nuclear transmutation} \end{cases}$$

where X is the nucleus of an isotope belonging to an element that composes the sample, p is the bombarding nuclear projectile, X* represents X in an excited state, Y and Z are residual nuclei of different elements, and b and c are ejected particles of different kinds. Nuclear transmutations are very important for activation analysis; the other two processes, viz., elastic and inelastic scatterings, are not that important. It may now be worthwhile to ask how many kinds of nuclear transmutations are possible in a certain set up? The answer is obtained by looking into the *Q*-value table. It depends upon X, p and energy of p.

Consider any particular nuclear transmutation, say

$$X+a \rightarrow Y+b \tag{7.10}$$

We know that in a nuclear transmutation, the mass and the charge numbers are conserved along with some other quantities. Therefore, X can be predicted if p, b and Y are known. Neutron capture reaction $(X+n \rightarrow Y+\gamma)$ is extensively used in activation analysis. In this transmutation, p is a thermal neutron, b is a gamma-ray and Y may be any nucleus depending upon X. X is not known. The main problem of activation analysis is to identify and measure Y, the residual nuclei. With high energy projectiles more than one type of nuclear transmutation may take place. This gives rise to interference, which may be taken into account in the analysis. The method of taking this into account has been discussed in a later section entitled 'Interferences'.

Sources of Activation. Neutron flux is the best source of activation, because even thermal neutrons can cause nuclear transformation, whereas high energy charged particles and gamma-rays are required to induce nuclear transmutation. In fact, thermal neutrons are better than fast neutrons. With thermal neutrons usually only capture reaction (n, γ) is possible. The cross-section of this reaction for heavier nuclei is usually larger. Therefore, with thermal neutrons analysis may be performed with less interference and high sensitivity, especially for heavier elements. With fast neutrons, many reactions (n, p), (n, α), $(n, 2n)$, (n, d), etc., may be possible, resulting in complicated interferences.

Thermal Neutrons. At present reactors are the best sources of high flux thermal neutrons. TRIGA reactors are commercially available. They operate at power levels in the range of 10-1000 kW. and produce usable fluxes in the range of 10^{11}-10^{13} n/cm^2 sec. They can be pulsed and are equipped with a rapid transfer pneumatic tube. They are pulsed, when very short-lived ($\sim 10^{-3}$/sec) activities are to be analysed. The pneumatic tube is used for activating samples, where the induced activity of interest has a half-life in the range of seconds to minutes. The use of reactors is limited to big research establishments because of their high cost.

Fast Neutrons. Fast neutron activation is essential for the trace analysis of light elements, because (n, γ) cross-section for these elements is small. They are detected with high sensitivity, using high cross-section (n, p), (n, α) and $(n, 2n)$ reactions on them. Fast neutron activation has gained more importance because of the availability of low cost ($\sim$\$24,000), high flux ($\sim 10^{10}$ n/cm^2 sec.) 14 MeV, neutron generators. The main drawback of these generators is that their flux is not steady.

Charged Particle Activation. Charged particles (proton, ^{3}He) have also been used as the source of activation. Proton-induced reactions are usually of high negative Q-value reactions. ^{3}He ions undergo a number of reactions with light elements in which energy is released (positive Q-value) ; hence, little or no additional energy is required to bring about these nuclear reactions, once the particle has penetrated the Coulomb barrier of the sample nucleus. But even a 8 MeV ^{3}He ion will penetrate only a few mils into the surface of a sample. This major limitation of all the charged particle activation, that the reaction permits only a surface analysis, is advantageous in some instances, e.g. metal surface studies.

Gamma-ray Activation. This kind of source has limited but selective applications. It is used for the determination of Be and ^{2}H by the following reactions:

$$^9\text{Be}+\gamma \rightarrow {}^8\text{Be}+n-1.66\text{MeV}$$

$$^2\text{H}+\gamma \rightarrow {}^1\text{H}+n-2.226\text{ MeV}$$

It is also used for the selective excitation of isomeric levels.

High energy bremsstrahlung (continuous hard X-rays) produced in electron accelerators are also used for activation with only limited applications.

Isotopic neutron sources, based on (α, n) and (γ, n) reactions, have also been used for the activation analysis. But the sensitivity with such sources is less, because of their low flux ($\sim 10^6$ n/sec. cm^2) Transuranic element californium-252 emits 10^9 n/sec. mg. and is, therefore a very good source of activation.

Alpha emitter curium-242 too is being used to activate characteristic X-rays. But it has very low penetration and generates heat in the sample.

Identification and Measurement of Activity. In a nuclear transmutation the residual nuclei, Y, are usually radioactive and formed in the excited states. They come to the ground state by emitting prompt gamma-rays (in a time 10^{-15} sec). Residual nuclei can be identified by recording either their characteristic prompt gamma-ray spectrum or radioactivity. The former method suffers from the following serious disadvantages: (1) details of the prompt gamma-ray spectra are not yet known, (2) these spectra are often too complicated and, therefore, difficult to interpret (3) the counting is done, while the irradiation is on; therefore, the background is large. This reduces the sensitivity.

The advantages of this method are: (1) analysis is always non-destructive and rapid, (2) analysis can be performed even if the residual nucleus is not radioactive, and (3) wide range of gamma-rays are available for the analysis.

The latter method based on radioactivity measurements does not suffer from the serious disadvantages of the former method, and is, therefore, normally used in actual practice.

The processes by which radionuclides decay are quite varied. The processes of maximum interest from the point of view of activation analysis are those involving beta (electron) and gamma (photon) emissions. The electrons radiated from a given nuclide are not monoenergetic but exhibit a continuous energy distribution from zero to some maximum energy, characteristic of the radionuclide. Photon emissions are monoenergetic and give rise to peaks in the gamma-ray spectrum. The energies of these photons are characteristic of the radionuclide producing them. The radionuclides may be identified by either beta or gamma counting. The latter method is superior because of the following reasons: (1) it offers double check of peak position and half-life, on the activity under measurements, (2) source self-absorption and scattering are small, therefore, comparatively larger samples can be counted and the sensitivity of the analysis can be enhanced, (3) many radionuclides present in the same specimen may be identified without chemical separation, even if they have very nearly equal half-lives.

In case the irradiated sample is thin and contains only two or three activities, beta counting with Geiger-Müller counter may be adopted for its analysis. However, the gamma counting with NaI (Tl) scintillation spectrometer is adopted if the irradiated sample is thick and contains many activities. The limit on the number of elements composing the sample that may be analysed without chemical separation (non-destructive) is imposed by the resolution of the spectrometer. NaI(Tl) gamma-ray spectrometer has about 8 percent resolution for 0.662 MeV gamma-ray. Li-Ge detector has less than $\frac{1}{2}$ percent resolution for this gamma-ray. Therefore, by using this detector, it is possible to analyse many more elements than with NaI (Tl) detector (without chemical separation). But the detection efficiency of the Li-Ge detector is low and it has to be operated at liquid air temperature in conjunction with low noise pre-amplifier. The detection efficiency may be increased by using large volume Li-Ge detectors that are now becoming available.

Even with the use of Li-Ge detector, it may not be possible to measure a very weak activity in the presence of a strong activity without chemical separation, especially if the characteristic gamma energy of weak activity is less than that of strong activity, because the full energy peak of the weak activity will be masked by the Compton radiation of the strong activity. However, this limitation may be overcome by such an anticoincidence counting which suppresses the Compton part of the gamma-ray spectrum. This may be done by surrounding the Li-Ge detector by NaI (Tl) crystal and recording those events that produce electrical pulses in Li-Ge detector only.

At present NaI(TI) detectors are being used in most of the activation analysis work, because of the non-availability of large size Li-Ge detectors. Here the Compton part of the gamma-ray spectrum is suppressed by putting NaI crystal in a well type CsI crystal. The source is placed on the top of NaI crystal. Both the crystals are coupled to the same photomultiplier. NaI pulses are faster than those of CsI. Therefore, if a photon loses its full energy in NaI only, the resulting pulse is faster. But if it loses a part of its energy in NaI and other part in CsI, the composite pulse is slower. The electronics is so arranged that it rejects the slower composite pulses and records only the faster, full energy pulses. In this way scattered radiation is rejected.

Radiochemical Separation. It is desirable that no radiochemical separation is performed in activation analysis, or else it is not non-destructive. But with samples containing many elements, the number of radioactive products may be so large that the activity measurement of a single component becomes almost impossible. In such cases, each activity or a group of activities may be chemically separated and measured. The resolution of the measuring set-up and the ratio of weaker to stronger product activities dictate the need for chemical separation. The lower is this ratio and resolution, the greater is the need for chemical separation.

Chemical separation may be performed before activation (pre-irradiation separation) or after activation (post-irradiation separation). The latter method is mostly adopted because no reagent blank is involved in it. There are many instances where the reagents contain an appreciable quantity of the elements to be determined. If the same reagents are used in pre-irradiation separation, the sought element in both the sample and reagents would become activated and subsequently measured, giving wrong information. A post-irradiation separation completely avoids this problem.

The choice of a separation method depends upon the physical, chemical and nuclear properties of the sought element. Various conventional chemical techniques like precipitation, solvent extraction, distillation, ion exchange, electrodeposition, volatilisation, etc., are often employed. The usual difficulties of micro chemistry are eliminated in radiochemical procedure. The loss of activity in the separation process is easily corrected. The two most widely used methods of yield correction are: (1) the addition of isotopic carrier to the sample, (2) the addition of radiotracer to the sample.

In the first method, a known weight ($\sim$10 mg) of stable atoms of the same chemical species as the radioactive atoms is added to the active sample before separation. If this weight is very large compared to the weight of the sought atoms in the sample, the following type of calculation can be used to determine the activity before separation.

$$A = A' \frac{W}{W'} \tag{7.11}$$

where A is the activity before separation, A' is the recovered activity, W is the weight added, and W' is the weight recovered.

The use of a radioisotopic tracer in a system is possible when a radionuclide of the sought element is available which is different from the activation analysis product. This is illustrated by the following example. In analysis of iodine, the ^{127}I isotope undergoes the reaction:

$$^{127}I + n \rightarrow {}^{128}I + \gamma$$

^{128}I decays with a half-life of 25 min. If before the separation of the ^{128}I activity from the sample, a known amount of ^{131}I (half-life 8 days) is added, the following relation is true:

$$A = A' \frac{A_1}{A_2} \tag{7.12}$$

where A is the activity of ^{128}I before separation, A' is the activity of ^{128}I recovered, A_1 is the activity of ^{131}I added, and A_2 is the activity of ^{131}I recovered.

Quantitative Analysis. The elemental composition of a sample can be qualitatively determined by examining the gamma-ray spectrum (peak position) and the decay curve (half-lives) of the irradiated sample. The quantitative analysis is performed with the help of the following expression:

$$D = \sigma \phi N (1 - e^{-\lambda t}) \tag{7.13}$$

where D is the rate of disintegration of the characteristic radioactivity just at the end of irradiation, N is the number of nuclides of the particular stable isotope of an element that produces this activity (we want to determine this number), σ is the cross-section for the reaction in which this activity is produced, ϕ is the flux of the nuclear projectiles that bombard the sample, $\lambda(=0.693/\text{half-life})$ is the decay constant of the activity, and t is the duration of irradiation.

The rate of disintegration D is connected to the counting rate by the following equation:

$$D = \frac{A(1+\alpha)}{\theta PC} \tag{7.14}$$

where A is the counting rate in the full energy peak of the selected gamma-ray, α, θ and P are the internal conversion coefficient, branching ratio, and the effective full energy peak efficiency of this gamma-ray respectively, and C is the source self-absorption and scattering factor.

Quantities σ, λ, α, θ, P and C occurring in equations (7.13) and (7.14) can be read from standard tables, charts and sheets. The values of A, t and ϕ, are measured in each irradiation. Knowing these quantities N can be calculated which may be converted into grams by the following expression:

$$m = \frac{N\,W}{N_0 K} \tag{7.15}$$

where m is mass of the element in grams, W is atomic weight of the element, N_0 is Avogadro's number and K is the fractional isotopic abundance of the given sample nuclide.

Absolute Sensitivity Calculation. The absolute sensitivity has been defined in many ways. The various definitions differ from each other on the amount of induced activity. According to one of them, the absolute sensitivity is equal to that mass of element which on being irradiated gives rise to maximum counting rate, which is at least equal to the background of the counting set-up. The term $(1-e^{-\lambda t})$ in equation (7.13) is called the saturation factor. As the time of irradiation (t) becomes large compared to the half-life of the product nuclide, the saturation factor approaches unity. Therefore, at the saturation time, the counting rate is maximum and is given by:

$$A = \frac{\theta PC}{(1+\alpha)} N\sigma\phi \tag{7.16}$$

According to the above definition of absolute sensitivity, the maximum counting rate should be at least equal to the background of the counting set-up. We notice from equation (7.16) that it depends mainly upon the cross-section of the reaction, the flux, the background and the efficiency of detection. It is not very difficult to build a gamma-ray counting set-up with high detection efficiency and low total background of about 2 counts/sec. One may put some reasonable values of the different quantities (read from nuclear tables and sheets), and calculate the minimum value of N that may give rise to actual counting rate of the order of the background. This is illustrated below by taking the following example:

$$\text{Background} = 2 \text{ counts/sec}$$

$$\frac{\theta PC}{(1+\alpha)} = 0.2$$

$$\sigma = 10^{-24} \text{ cm}^2$$

$$\phi = 10^{12} \text{ neutrons/cm}^2 \text{ sec}$$

Then

$$2 = 0.2 \times 10^{-24} \times 10^{12} \times N$$

$$N = 10^{13} \text{ atoms}$$

If these are aluminium atoms, which are monoisotopic in nature ($K=1$; $W=26.98$), then the minimum mass of this element, according to equation (7.15), that can be detected is

$$m = \frac{10^{13} \times 26.98}{6.02 \times 10^{23}} = 0.00045 \times 10^{-6} = 0.00045 \,\mu\text{g}$$

Parts Per Million Sensitivity. The parts per million (p. p. m.) sensitivity is defined in the following way:

$$\text{p.p.m. sensitivity} = \frac{\text{absolute sensitivity in } \mu\text{g}}{\text{mass of the sample in g}}$$

Therefore, if a sample of 10 g can be irradiated and counted, then according to the above calculations the minimum mass of the element that can be detected per gram of the sample is equal to

$$\frac{0.00045}{10} = 0.000045 \,\mu g$$

That is, the p.p.m. sensitivity of the activation analysis in this case is 0.000045. It should be emphasised that calculated sensitivities can be very misleading, since interferences by other activities present may cause the actual attainable sensitivities to be much poorer than those predicted. Nevertheless, calculated values may usefully serve as a guideline.

Errors. The accuracy of the various quantities that enter in the calculation of N is such that the overall accuracy of the absolute activation analysis is not better than ± 15 per cent. The maximum error is on the cross-section ($\sim\pm 12$ per cent). Therefore, there is a need to measure cross-sections with better accuracy. In fact, due to large errors in the nuclear parameters required for the calculations of N, virtually all activation analysis work is done by the comparison method. In this method, each activity induced in the sample is compared with similar activity, induced in the standard of similar size and composition irradiated exactly in the same way as the sample. Because of this comparison all the nuclear parameters get cancelled and the accuracy of the analysis is improved by 1 to 3 per cent. The expression for m [equation (7.15)] then takes the following form:

$$m = \frac{\text{activity of the sample} \times \text{mass of the standard}}{\text{activity of the standard}}$$

But then the versatility of the technique is reduced in the sense that one will have to know the element one is looking for. Further, this method involves more time and work also, specially if many elements are to be analysed.

Interferences. There are two main types of interferences: (1) overlapping of two or more photopeaks, and (2) production of same activity from the activation of different elements or isotopes. The first one occurs whenever one or more induced gamma-ray emitters emit gamma-rays, of nearly the same energy as that of the induced activity of interest, in statistically significant amounts and where decay does not eliminate them. The closer the two gamma-rays are to one another in energy and the greater the counting rate of the interfering activity relative to the one of interest, the greater is the extent of interference. Often such interfences can be minimised by proper choice of irradiation and decay times, and by using a detector of better energy resolution. Also, a pure photopeak of interest can be obtained by the process of spectrum stripping, that is, subtracting the undesired gamma-ray from the sample spectrum. Alternatively, the interfering contribution can be removed mathematically by computer calculation. If the interfering peak is due to activity of a different element, it can be removed by chemical separation.

The second type of interference occurs in high energy activation of samples, containing neighbouring elements. For example, ^{41}Ar is formed both in the activation of calcium and potassium by the 14 MeV neutrons from $^{44}Ca(n, \alpha)^{41}Ar$ and $^{41}K(n, p)^{41}Ar$ reactions. Therefore, from observed ^{41}Ar activity it is not possible to say whether both or only one element is present in the sample. In such cases, one should tabulate other possible reactions (n, α), (n, p) and $(n, 2n)$ on these isotopes (in this case on ^{44}Ca and ^{41}K), the end products of which are radioactive. In the present example they are:

$$^{44}Ca(n, p)^{44}K \text{ and } ^{41}K(n, \alpha)^{38}Cl$$

Hence one should look for ^{44}K and ^{38}Cl activities in the irradiated sample in order to remove the ambiguity. If both these activities are present, the sample contains both calcium and potassium. But if only one, say ^{44}K, is present then the sample contains calcium and not potassium.

But if no activation product of potassium other than ^{41}Ar was radioactive (i.e. if ^{38}Cl was stable), then the content of ^{44}Ca would be calculated from ^{44}K activity, and the contribution of (n, α) reaction on this much amount of ^{44}Ca in observed ^{41}Ar activity would be determined. The excess activity of ^{41}Ar if any, would be because of the activation of potassium.

Assessment of the Technique. The steps to be followed to establish the feasibility of activation analysis in a given problem are detailed in Fig. 7.2. It may be noticed that the principal limitations of the activation technique arise from the nuclear properties of an element and not its chemistry.

Activation analysis is an expensive technique. Initial set-up may cost as much as $ 60,000, if one uses a 14 MeV, high flux neutron generator. However, because of its high speed, this technique may turn out to be not so expensive in the long run, provided a large

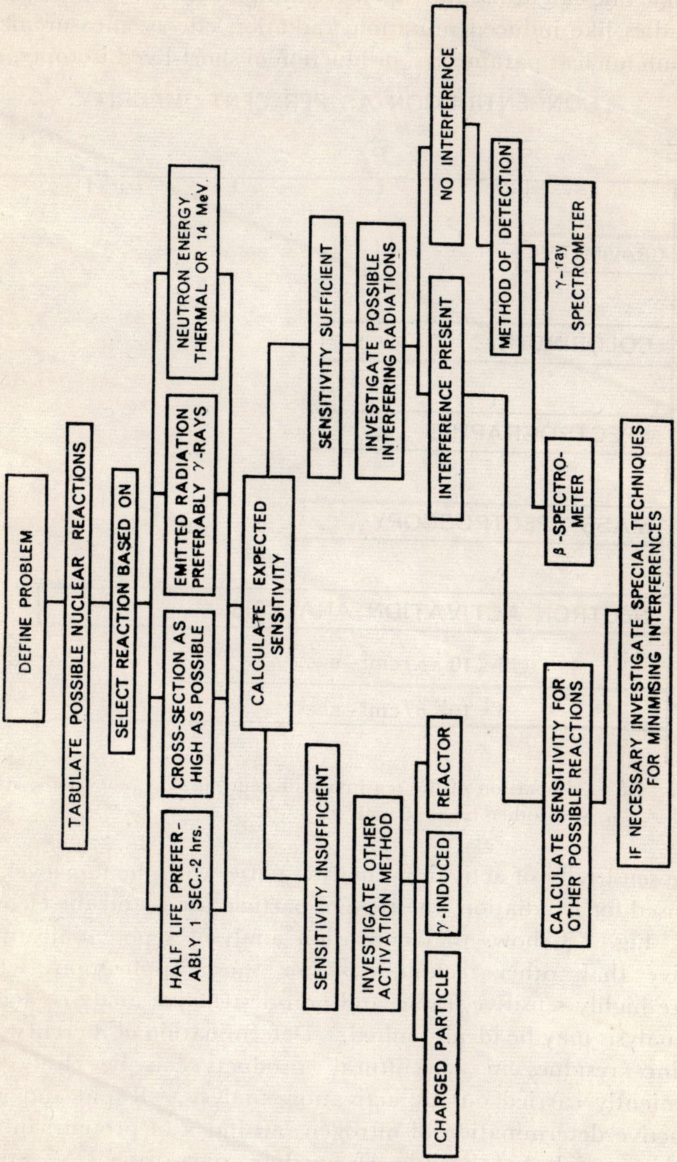

Fig. 7.2 Guide chart for establishing the feasibility of activation analysis.

number of samples are to be handled. In case the bulk of work is not large, one can utilise the expensive components for other interesting studies like induced mutation, radiation effects, measurements of certain nuclear parameters, production of short-lived isotopes, etc.

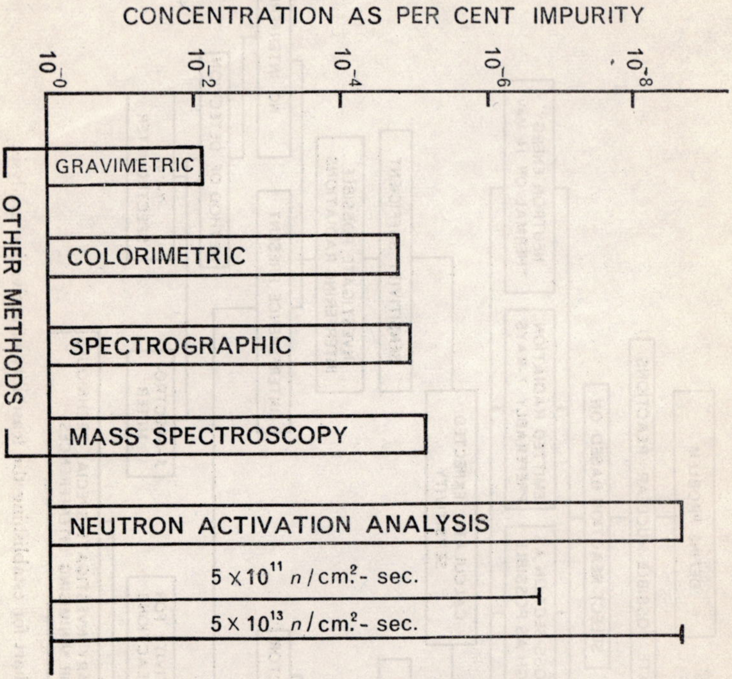

Fig. 7.3 Comparison of the sensitivity of neutron activation analysis with other types of analyses.

The sensitivity of activation analysis varies with the flux level, the time used for irradiation, the type of particle used, and the element itself. Fig. 7.3 shows that activation analysis is potentially more sensitive than other techniques. For those applications which require highly sensitive, rapid and non-destructive analysis, activation analysis may be ideally suited. Determination of mercury and bromine residues in agricultural products can be easily and conveniently carried out by activation analysis. Rapid and non-destructive determination of nitrogen (an index of protein) in biological materials has been performed by measuring ^{13}N activity produced by 14 MeV neutrons in the following reaction:

$$^{14}N + n \rightarrow {}^{13}N + 2n$$

The 10.82 MeV prompt gamma-ray produced by thermal neutrons in $^{14}N(n, \gamma)^{15}N$ reaction has also been utilised for the measurement of nitrogen in organic materials. It may be noted that no other element in the entire periodic table emits such a high energy (10.82 MeV) gamma-ray in (n, γ) reaction. It is probable to use this gamma-ray for rapid and non-destructive determination of protein in single wheat seed, provided a pure thermal neutron beam having a flux of about 10^9 *n*/sec cm^2 is availbale. It may be possible to measure the protein content of 10 g wheat sample in 1 min, if a pure thermal neutron flux of about 10^6 *n*/sec cm^2 is available.

7.4.2 Neutron Moisture Meter

The neutron moisture meter is an instrument for rapid, reliable and non-destructive measurement of soil moisture. This instrument has opened the way for intensive studies of water use efficiency. These studies are prerequisites for developing more efficient methods of water use which are extremely important for producing more food from less water. Although, the importance of these studies has been well-known for a long time, very little work could be done because a large number of soil moisture measurements at various depths are required, for which no suitable method was known before the neutron moisture meter came into the picture.

Construction. The neutron moisture meter has three main parts: a fast neutron source of low intensity, a detector of slow neutrons and electronic units (power supply, preamplifier and scaler) for the detector. Both the fast neutron source and detector of slow neutrons are enclosed in a water tight tube called probe. Its length is usually 35 cm, and diameter 4 cm. Fig. 7.4 shows the various parts of a neutron moisture meter. The fast neutron source is usually a mixture of radium and finely ground beryllium. It is called Ra-Be source. In this source about 5 MeV alpha-particles one emitted by radium which produce fast neutrons in their interaction with beryllium. The nuclear interactions that lead to the emission of fast neutrons from Ra-Be source are as follows:

$$Ra \rightarrow {}^{282}_{86}Rn + He + \gamma$$
$$He + Be \rightarrow C + n \quad \text{(fast neutrons)}$$

The half life of ^{226}Ra is more than 1500 years. Therefore, Ra-Be mixture works as a source of constant neutron flux for a long

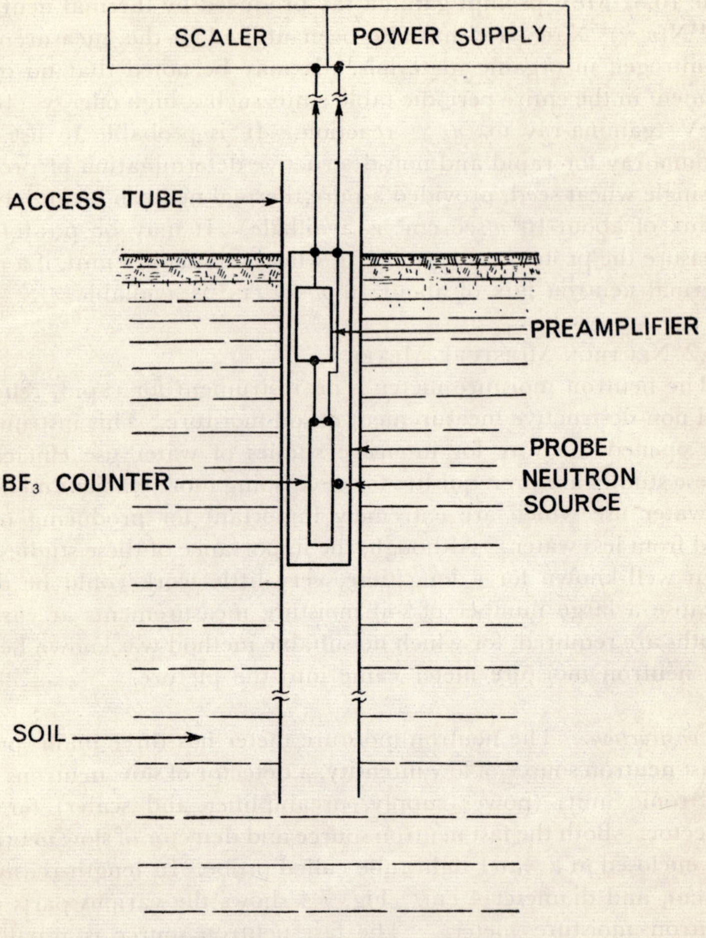

Fig. 7.4 Neutron moisture meter in soil.

time. A few mC of ^{226}Ra is used in neutron moisture meter.

The detector is usually proportional counter, filled with ^{10}B enriched BF_3 gas. When a slow neutron enters this detector, alpha-particle is produced in it due to the following nuclear reaction:

$$^{10}_{5}B + ^{0}_{0}n \text{ (slow)} \rightarrow ^{7}_{3}Li + ^{4}_{2}He \quad \text{(alpha-particle)}$$

The alpha-particle produced in the detector causes ionisation in it, which leads to the generation of electrical pulse that is recorded by the scaler. It is important to note that this detector cannot detect the fast neutrons that are emitted by Ra-Be neutron source. It detects slow (thermal) neutrons with very high efficiency. As a consequence of this, the counting rate is very low if there is no slowing down material (hydrogen) in the vicinity of the neutron moisture meter probe.

Basis of Measurement. The measurement of the moisture is based on the fact that the fast neutrons which are emitted by the source lose energy by colliding with the nuclei of the surrounding material. The process of energy loss continues until the original fast neutron has been completely slowed down. Such neutrons are called thermal neutrons. While the neutron is losing energy, its motion is becoming random and it starts moving in all directions. The degree of randomness increases as the neutron loses more and more energy. The motion of a thermal neutron is as random as the motion of a gas molecule. It is only these randomly moving thermal neutrons that are counted by the neutron moisture meter.

As explained in section 4.4.1, the neutron suffers greatest energy loss in collision with hydrogen nuclei. That is, hydrogen is the best medium for slowing down (moderation) of fast neutrons. Heavier elements have poor moderating capacity and contribute very little in the counting rate of neutron moisture meter. The contribution of lighter elements, especially of hydrogen is the maximum.

In soil, hydrogen is present mostly in the form of water. Therefore, the counting rate of a neutron moisture meter, whose probe is inserted in soil, represents a measure of moisture content of soil. The counting rate is converted into the percentage of moisture with the help of a calibration graph prepared by using the soil of known moisture. Elements other than hydrogen that are efficient moderators of neutrons are lithium, boron, chlorine and manganese. But fortunately, in soils they are present only in very small quantities as compared to many kg of water per m^3 of soil of normal moisture content. The presence of these elements in concentration of the order of a gram per m^3, have no influence on the moisture measurement. However, the presence of strong thermal neutron absorbers like ^{10}B in small amounts influences the moisture measurements.

Advantages and Disadvantages. The moisture measurement with neutron moisture meter is rapid; usually 5 minutes are enough for a single measurement. The accuracy of the measurement is satisfactory. The probe gives moisture content of an almost spherical volume of soil ranging from 30 to 75 cm in diameter, depending inversely upon the moisture content of soil. The measurements are non-destructive, allowing for long-term studies on the same soil volume.

The main disadvantages of the neutron moisture meter are : (1) high cost, (2) rather heavy weight, (3) desired precision is not achieved when the moisture content is low, because the sphere of influence is then large, (4) a special probe is required to measure the moisture content of the top (30 cm) soil layer, and (5) a good electronic technician is required to repair and maintain the instrument. In spite of these disadvantages, the neutron moisture meter is solving the problem of performing a large number of reliable soil moisture measurements, which are essential to develop efficient methods of water use.

7.4.3 Seed Oil Mass Screening by NMR Technique

The major aim of oil crop breeding programmes is to produce more oil per day per hectare. The total quantity of oil produced by an oil crop is a product of its yield and percentage of oil. The breeder uses visual methods for evaluation of high yields and other characters like, resistance to diseases, pest and arid conditions, the ability of the crop to withstand wind and rain and its ability to reach early maturity. But he faces enormous difficulties when selecting seeds and plants for higher percentage of oil among thousands of progenies on the basis of slow chemical analysis. Further, chemical analysis is destructive and does not permit direct propagation of seeds. NMR (nuclear magnetic resonance) technique is used for rapid, non-destructive and reliable determination of oil in oil seeds.

Basic Principle. Atomic nuclei like, ^{1}H, ^{19}F, ^{31}P, etc. with non-zero nuclear spin possess magnetic moments. If an external magnetic field is applied to a system of such nuclei, these nuclear magnets will experience torques and will tend to become parallel to the magnetic field. A bar magnet placed in a magnetic field does the same thing. But there is a basic difference between the two. A bar magnet does not spin, whereas a nuclear magnet spins. As a result of this, the nuclear magnet placed in magnetic field precessess about the field. This motion is similar to that of a gyroscope.

The frequency of precession depends upon the spin and magnetic moment of the nucleus and the strength of the applied magnetic field. If a radiofrequency magnetic field is applied on the precessing magnetic nuclei in an appropriate way, these nuclei will absorb energy if the frequency of precession is equal to the frequency of radiofrequency field. This is called nuclear magnetic resonance (NMR). It gives rise to absorption peaks in the NMR spectra. The width of the peaks depends very strongly upon the mobility of the magnetic nuclei in the sample. Higher the mobility, narrower is the peak.

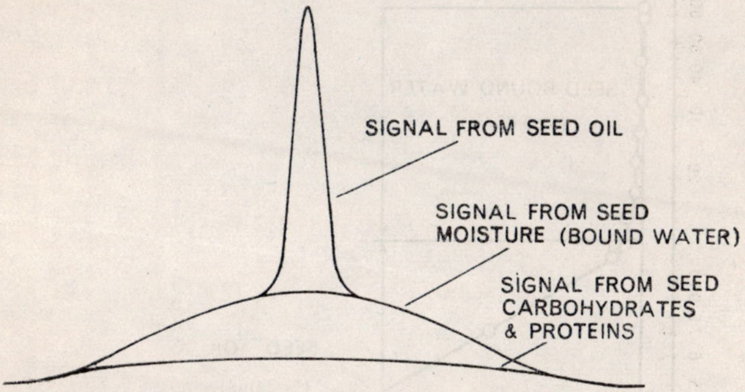

Fig. 7.5 Wide-line NMR signal from seed oil, seed moisture, and seed carbohydrates and proteins.

In oilseeds, hydrogen which gives strong NMR signal is present mainly in four forms, namely, oil, moisture, carbohydrates and proteins. The oil hydrogen is most mobile among them. Therefore, its NMR signal (peak) is narrowest, as shown in Fig. 7.5. It may be noted that NMR signals due to bound water (moisture) and solid hydrogen (seed protein and carbohydrate) are practically constant in the region of seed oil peak. This makes it possible to determine the height of the oil peak without any significant interference from other peaks. The height is determined electronically and is proportional to the quantity of oil in seeds. The constant of proportionality is determined by means of a calibration graph drawn by using the oil of the same crop. This method of oil measurement is quite fast and non-destructive, and is called Wide-Line NMR technique for seed oil determination.

Recently, the author and his group have developed Pulsed NMR technique for similar analysis. The basic difference between Wide-Line NMR and Pulsed NMR is that in the former a continuous radiofrequency magnetic field is applied on the sample placed in a uniform magnetic field, whereas in the latter, a discontinuous (pulsed) radiofrequency magnetic field is applied on the sample. The effect of radio frequency magnetic field applied to the sample at resonance frequency is to tip the magnetisation vector away from its equilibrium direction. The pulse which can turn the magnetisation vector perpendicular to the uniform magnetic field is referred to as 90° or $\pi/2$ pulse.

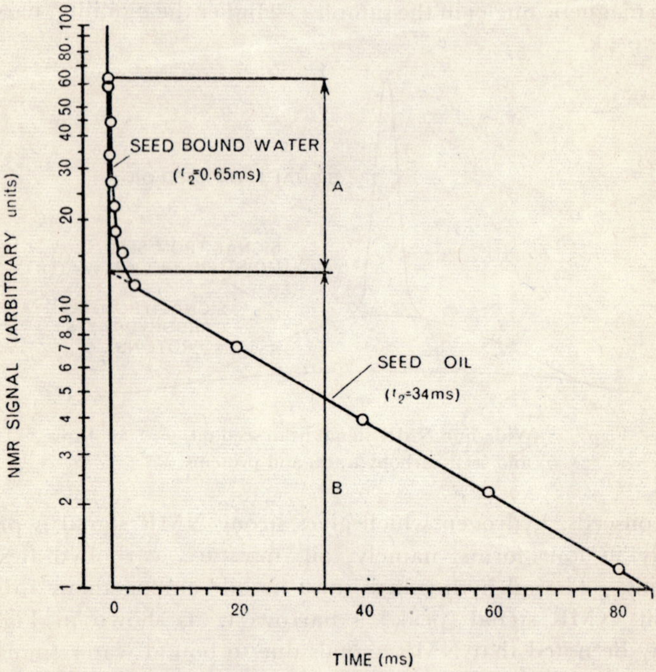

Fig. 7.6 Pulsed NMR signal after 90° pulse from seed moisture and seed oil.

After the end of a 90° pulse, the magnetisation vector decays exponentially, being proportional to $\exp(-t/T_2)$; T_2 is the spin-spin or transverse relaxation time and t is the time from the end of the pulse. T_2 depends very strongly on the mobility of the magnetic nucleus under observation. It increases with increase in mobility of the nucleus. T_2 of seed oil hydrogen is higher than T_2 of other forms of seed hydrogen. This is because seed oil is more mobile than any other constituent of the seed. As a consequence of this, the pulsed NMR signal from oil persists as shown in Fig. 7.6, upto a time when all other signals have decayed. This makes it possible to measure the oil signal without any interference from other components of the seed. The measured signal is converted into quality of oil with the help of a calibration graph prepared using pressed oil from the same crop. The use of such a calibration is based on the fact that the variation in hydrogen percentage of oil from different varieties of the same crop is found to be not more than 2%; of course, an error to this extent is introduced in the analysis. The analysis is highly reproducible, rapid and non-destructive. Measurements of NMR signal takes about half a minute. Practically, no sample preparation is required, except that the seeds should be clean and dry. An important advantage of this technique over the Wide-Line NMR technique is that its requirement of magnetic field homogeneity is not critical.

Bibliography

ABRAGAM, A., (1961), *Principles of Nuclear Magnetism,* Oxford University Press, London.

VIII International Symposium of Agrochimica on the Application of Nuclear Energy in Agriculture, (1971), *Agrochimica* **15,** 2-3.

ARONOFF, S. (1961), *Techniques of Radio-biochemistry,* Iowa State University Press, Ames.

BELL, G. A. and F. N. HAYES (eds., 1958), *Liquid Scintillation Counting,* Pergamon Press, Oxford.

BIRKS, J. B. (1953), *Scintillation Counters,* Pergamon Press, Oxford.

BOURSNELL, J. C. (1958), *Safety Techniques for Radioactive Tracers,* Cambridge University Press, Cambridge.

BOYD, G. A. (1955), *Autoradiography in Biology and Medicine,* Academic Press, New York.

BURCHAM, W. E. (1963), *Nuclear Physics, An Introduction,* Longmans Green, London.

CASARETT, A. P. (1968), *Radiation Biology,* Prentice-Hall, Englewood Cliffs, N. J.

CHASE, GRAFTON D. and JOSEPH RABINOWITZ, (1962), *Principles of Radio-Isotope Methodology,* Burgess Publishing Co., Minneapolis.

COMAR, C. L. (1955), *Radioisotopes in Biology and Agriculture,* McGraw-Hill, New York.

CONWAY, T. F. and F. R. EARLE (1963), NMR for Determining Oil Content of Seeds, *Journal of the American Oil Chemists' Society* **40,** 265.

CORNWELL, P. B. (ed., 1966), *Entomology of Radiation Disinfestation of Grain,* Pergamon Press, Oxford.

DEARNALEY, G. and D. C. NORTHROP (1966), *Semiconductor Counters for Nuclear Radiation,* 2nd ed., E. & F. N. Spon, London.

DUIVENSTIJN, A. J. and L. A. VENERLOO (1963), *Practical Gamma Spectrometry,* C. C. Thomas, Springfield.

DZHELEPOV, B. S. and L. K. PEKER (1961), *Decay Schemes of Radioactive Nuclei,* Pergamon Press, Oxford.

ENGE, HERALD A. (1966), *Introduction to Nuclear Physics,* Addison-Wesley, Mass.

EVANS, R. D. (1955), *The Atomic Nucleus,* McGraw-Hill, New York.

FAO/IAEA (1964), Laboratory Training Manual in the Use of Isotopes in Soil-Plant Relations Research, *Technical Report Series* **29,** Vienna.

——(1966), Proceedings of International Symposium on Radiation Preservation of Foods, *STI/PUB* **127,** Vienna.

——(1969), *Proceedings of the Symposium on the Induced Mutation in Plants,* Vienna.

——(1967), *Proceedings of the Symposium on Isotopes in Plant Nutrition and Physiology,* Vienna.

FISHER, A. (1930), *Statistical Method for Research Workers,* Oliver and Boyd, Edinburgh.

FREMLIN, J. H. (1964), *Application of Nuclear Physics,* English University Press, London.

FRIEDLANDER, G., J. W. KENNEDY, and J. M. MILLER (1964), *Nuclear and Radiochemistry,* 2nd ed., John Wiley & Sons, New York.

GAMOW, GEORGE, and J. M. CLEYELAND (1961), *Physics, Foundations and Frontiers,* Prentice-Hall, Englewood Cliffs, N. J.

GLASSER, ALVIN (1961), *Introduction to Nuclear Science,* D. Van Nostrand, New York.

GLASSTONE, SAMUEL, (1967), *Sourcebook on Atomic Energy,* D. Van Nostrand, New York.

GUDE, W. D. (1968), *Autoradiographical Techniques,* Prentice-Hall, Englewood Cliffs, N.J.

HARVEY, BARNARD G. (1962), *Introduction to Nuclear Physics and Chemistry,* Prentice-Hall, Englewood Cliffs, N.J.

HENRY, HUGH F. (1969), *Fundamentals of Radiation Protection,* John Wiley & Sons, New York.

HOLLANDER, J. M. and I. PERLMAN (1966), The Semiconductor Revolution in Nuclear Radiation Counting, *Science* **154,** 84-93.

IAEA (1962), *Applications of Isotope Techniques in Hydrology,* Vienna.

——(1966), Laboratory Training Manual on the Use of Isotopes and Radiation in Entomology, *Technical Reports Series* **61,** Vienna.

——(1970), Manual on Mutation Breeding, *Technical Report Series* **119,** Vienna.

——(1967), *Proceedings of Symposium on Isotope and Radiation Techniques in Soil Physics and Irrigation Studies,* Vienna.

——(1967), *Proceedings of Symposium on Nuclear Activation Technique in Life Sciences,* Vienna.

——(1968), *Proceedings of Symposium on Radiations in Soil Organic-matter Studies,* Vienna.

——(1962 & 1965), *Proceedings of Symposium on Radioisotopes in Soil-Plant Nutrition Studies,* Vienna.

——(1970), *Proceedings of Symposium on the Use of Isotopes in Hydrology,* Vienna.

——(1968), *Radiation, Radioisotopes and Rearing Methods in Control of Insect Pests,* Vienna.

KAMEN, M. D. (1957), *Isotopic Tracers in Biology,* Academic Press, New York.

MEYERHROF, WALTER E. (1967), *Elements of Nuclear Physics,* McGraw-Hill, New York.

National Bureau of Standards (1953), Maximum Permissible Doses of Radioisotopes, *Handbook Series* **52,** Washington.

OVERMAN, RALPH T., and HERBERT M. CLARK (1960), *Radioisotope Techniques,* McGraw-Hill, New York.

PARRAT, L. G. (1955), *Probability and Experimental Errors in Science*, John Wiley & Sons, New York.

PRICE, WILLIAM J. (1964), *Nuclear Radiation Detection*, McGraw-Hill, New York.

QUIMBY, E. H., and S. FEITALBERG (1962), *Radioactive Isotopes in Medicine and Biology*, Lea & Febiger, Philadelphia.

ROBALT, J. (1950), Photographic Emulsion Techniques, *Progress in Nuclear Physics*, V. 1 (Ed. O. R. Frisch), Pergamon Press, Oxford.

ROBOZ, JOHN (1968), *Introduction to Mass Spectrometry, Instrumentation and Techniques*, John Wiley & Sons, New York.

ROMER, A. (ed., 1973), *The Discovery of Radioactivity*, Dover, New York.

ROTHCHILD, S. (ed., 1963), *Advances in Tracer Methodology*, Plenum Press, New York.

SCHRAM, ERIC (1963), *Organic Scintillation Detector*, Elsevier, New York.

SHARPE, J. (1955), *Nuclear Radiation Detectors*, Methuen & Co., London.

SMITH, C. M. H. (1965), *A Textbook of Nuclear Physics*, Pergamon Press, Oxford.

THORNBURN, C. C. (1972), *Isotopes and Radiations in Biology*, Butterworth & Co., London.

TIWARI, P. N. (1971), Rapid and Non-Destructive Determination of Protein in Grain Samples by Prompt (n, γ) Technique, *Radiochem. Radioanal. Letters* **6**, 363.

TIWARI, P. N., P. N. GAMBHIR, and R. S. RAJAN (1974), Rapid and Nondestructive Determination of Seed Oil by Pulsed NMR Technique, *Journal of the American Oil Chemists' Society*, **51,** 104-109.

UNO (1955, 1958, 1969 and 1971), *Proceedings of the Peaceful Uses of Atomic Energy*, Geneva.

WANG, C. H. and D. L. WILLIS (1965), *Radiotracer Methodology in Biological Science*, Prentice-Hall, Englewood Cliffs, N.J.

WILSON , J. G. (1951), *The Principles of Cloud Chamber Technique*, Cambridge University Press, Cambridge.

WOLF, GEORGE (1969), *Isotopes in Biology*, Academic Press, New York.

Index